PLASTICISERS: PRINCIPLES AND PRACTICE

Plasticisers
Principles and Practice

Alan S. Wilson

The Institute of Materials

Book 585
First published in 1995 by
The Institute of Materials
1 Carlton House Terrace
London SW1Y 5DB

ISBN 0 901716 76 6

Typeset by Fakenham Photosetting Ltd
Printed and bound in the UK at
The University Press, Cambridge

Contents

CHAPTER 6   THE SELECTION PLASTICISERS FOR SPECIFIC
APPLICATIONS OF PVC

CHAPTER 7   PLASTICISERS FOR POLYMERS OTHER THAN PVC

CHAPTER 8   PLASTICISER QUALITY, SPECIFICATIONS AND ANALYSIS

Preface

The purpose of this book is to provide a concise set of information covering the various inter-related issues which are of interest to people whose work requires some knowledge of plasticisers. Whilst it is aimed principally at industrial users and suppliers of plastic additives it is hoped that the book will also be of value to any researcher or student who wishes to gain an appreciation of plasticisers in their industrial context.

The selection of contents is based on the range of questions personally encountered during thirty years of working with these materials in the fields of technical service, development and market research. These questions have come from customers and colleagues, processors and raw materials suppliers, experienced plasticiser users and newcomers. They relate not only to technology but to the commercial status and raw materials basis of plasticisers, to questions of quality and suitability for purpose, and in steadily increasing volume to the health, safety and environmental aspects of these important industrial chemicals. The last of these is a large dynamic subject and the coverage in the last chapter of the book is no more than a snapshot of the scene at the time of writing. It attempts to provide an integrated picture of a number of apparently diverse issues which have come to the fore in recent years.

Chapters 4–7 make reference in several places to the relative prices of different plasticisers, this being the major factor accounting for their long term commercial status. The writing of the book coincided with an unprecedented threefold escalation in European prices for large tonnage commodity phthalate plasticisers. So far these high prices have persisted until the middle of 1995 and this temporarily invalidates some of the comparisons made. However since they are based on more typical historical relationships they are likely to be reasonably representative of the situation if and when phthalate prices return to a stable level.

The science and technology of plasticisers and plasticisation continues to attract research effort, nowadays having the benefit of much more searching techniques than were available in the pioneering days. In contrast to HSE and economic issues it is generally seen as a mature subject with a low probability of major new developments. Hence purely technical texts on plasticisers tend not to date quickly. The milestone publication which above others is something of a bible for plasticiser specialists is *The Technology of Plasticisers* by Sears and Darby which draws on comprehensive research spanning the full range and history of the subject. However, the average plasticiser user in need of a quick answer to a technical question is likely to rely on suppliers' literature backed by access to their technical service representatives. The technical data presented and discussed in this present book are mainly of the type which is available from these sources. The intention is to aid the interpretation and application of such data and it is hoped that this will be of value in contacts between suppliers and users. The same theme occurs in Chapter 8 of the book which covers the subject of plasticiser quality and specifications.

The balance of contents of the book reflects the fact that the great majority of consumption of plasticisers is in PVC. This dominance sometimes causes other applications to be overlooked and Chapter 7 attempts to redress this by covering other polymers in which the use of plasticisers is technically and commercially important.

Thanks are due to the many companies and individuals who have contributed information both in the form of technical literature and of vital facts and opinions based on their experience. I am particularly grateful to BP Chemicals, the source of much of the reference data contained in the book and to Paul Brown, David Cadogan, Francis Carpanini, Roland Caulcutt, Paul Clutterbuck, Peter Danckwerts, Jennifer Hodge, David Pugh, Alex Thomson and my wife Lyn for their help and encouragement.

CHAPTER 1

Industrial Status, Function and Mechanism of Plasticisers

1.1   DEFINITION

A plasticiser may be described generally as a 'substance incorporated into a material to increase its flexibility, softness, distensibility or workability'. This book deals with additives used to modify the end properties of synthetic polymers, thereby making them useful materials for a wider range of applications. The German term 'Weichmacher' (softener) is perhaps more clearly descriptive of their function. In fulfilling the primary aim of achieving the required properties for the end product, plasticisers may also give benefits in easier forming and fabrication. In this secondary function, they are acting as **process aids**. In contrast to plasticisers, additives used specifically as process aids are usually required to facilitate processing without altering end properties.

Plasticisers are typically high boiling, oily, organic liquids. The great majority of the materials in commercial use are esters. This is not an intrinsic requirement but neither is this feature mere coincidence. More will be said on this in Chapter 5.

The process by which these materials function is described as **external plasticisation** since they are additives which do not form part of the polymer structure. **Internal plasticisation** refers to the structural modification of a polymer to reduce its glass transition temperature and make it more flexible at ambient temperature. This is best illustrated by reference to polyvinyl acetate (see Chapter 7, 7.2.1) where the two processes compete to some extent.

Since plasticisers are not chemically bound to the polymer, they are subject to loss by migration under service conditions. Whilst this is the main technical limitation of this whole area of technology, materials selected for commercial use must be of sufficiently high molecular weight to keep losses within acceptable limits during the design life of the end product.

Nowadays any search of the technical literature for the keyword 'plasticiser' produces a large number of references to concrete technology. Here the function of these additives is to increase the fluidity of cement pastes, thereby increasing the workability of concrete at a constant water/ cement ratio. Alternatively, they permit concrete to be made with lower water levels with a consequent increase in strength.[1] A newer generation of such additives with greatly increased efficiency is termed 'super-plasticisers'.[2] Any similarity between these concrete additives and the polymer modifiers covered by this book is tenuous.

1.2 STATUS OF PLASTICISERS IN THE PLASTICS INDUSTRY

Each year, western Europe uses about one million tonnes of plasticisers with a current (1994/5) value of the order of £1,000 M. This makes them the largest class of plastics additives both in terms of value and (if fillers are excluded) in volume. Table 1.1 shows the relative status of various

TABLE 1.1
1990 EC CONSUMPTION OF PLASTICS ADDITIVES
(by courtesy of Frost and Sullivan)

	('000 TONNES)	($MILLION)
Plasticisers	984	1 023
Flame Retardants	214	435
Antioxidants	25	163
Heat Stabilisers	88	264
UV Stabilisers	4	80
Degradability Additives	7	23
Pigments	277	658
Chemical Blowing Agents	8	29

types of plastics additives in the European Community in 1990.[3] The tonnage of plasticisers used annually exceeds that of any polymer class, apart from the highest volume thermoplastics. About 90% of plasticiser consumption is in PVC, the remainder being divided between polyvinyl acetate (adhesives), synthetic rubbers, polyurethane sealants, cellulosics and a variety of other polymers.

1.3 DEPENDENCE OF THE PVC INDUSTRY ON PLASTICISERS

PVC, for much of recent history the cheapest thermoplastic, is unique in its ability to accept plasticisers at high loadings and still give technically useful materials. For the newcomer to the subject of plasticisation, the levels of plasticiser used in everyday articles made from PVC are some-times a matter for astonishment. In a PVC Wellington boot, for example, a liquid plasticiser accounts for more than half the volume of the material present. Recognition of this attribute long ago led to the evolution of PVC as the most versatile of thermoplastics. It is within the capability of the small plastics processor with limited capital investment to tailor materials with a wide range of mechanical properties for different end uses from a single polymer and a single plasticiser.

A major factor accounting for the variety of end uses for plasticised PVC is the availability of grades of polymer capable of forming liquid dispersions in plasticiser. Polymer/plasticiser ratios in these dispersions (plastisols) can be adjusted to give a useful range of physical properties in the end product. Processing via the plastisol route gives an enormous range of application and design possibilities.

Whilst all flexible PVC compositions share common characteristics, selection of the plasticiser type used makes it possible by simple formulation to adapt them to a wide range of processes and end uses. Thousands of chemicals have suitable characteristics for use as plasticisers. The number of distinct chemicals in actual commercial use as plasticisers is sometimes overestimated but may be around 200. However, high price confines the majority of these to specialised niches. It is the favourable cost/performance of flexible PVC and the variety and control of properties offered to the formulator which have lead to its current status.

The development of commercial applications for PVC in the 1930s and 1940s depended on plasticisation and the use of unplasticised PVC did not become significant until much later. By the 1960s the technology for formulating and processing rigid PVC had made great strides and markets for rigid PVC products were growing at a faster rate than for flexibles. Long familiarity with the plasticised form of the polymer often gave the public the impression that rigid PVC articles were made from something new and completely different – UPVC. The standard designations **PVC-U** (unplasticised) and **PVC-P** (plasticised) have now been adopted by IUPAC for the two forms of the material. The applications of rigid PVC are less diverse than those for flexible PVC and are dominated by large tonnage extruded products for the construction market. The low cost and fire resistance of PVC have been major factors in its success here. The differential in growth rates between unplasticised and plasticised PVC has now resulted in rigid outlets accounting for around two thirds of consumption of the polymer.

1.4 THE MAJOR PLASTICISERS

Of the one million tonnes of plasticisers used annually in Europe, approximately 90% is phthalate esters. The great majority of phthalate consumption is of the 'big three' general purpose PVC plasticisers:

DOP   dioctyl phthalate
             (actually di 2–ethylhexyl phthalate)
DINP   di–isononyl phthalate
DIDP   di–isodecyl phthalate

The largest single product used as a plasticiser in Europe and throughout the world is DOP. Unusually for a product of this type, it consists of a single chemical component and this sometimes makes it an easier subject of study than many other plasticisers. DOP is universally accepted as the standard of comparison in the testing of PVC plasticisers. The subject of the phthalate esters is covered in detail in Chapter 4.

1.5   HISTORICAL REVIEW

At this point, it is instructive to look back through the various stages by which industrial plasticisers reached their current status. From the beginning of the twentieth century until about 1930, early plasticiser technologists were preoccupied with nitrocellulose. After 1930, when it was discovered how to process PVC, attention shifted to this polymer where it has remained ever since and nitrocellulose was relegated to a minor position. By the 1960s, the technology of PVC plasticisation had reached a mature stage and no really new plasticiser of any significance has been developed since then.

1901 to 1930: The nitrocellulose era
By the turn of the century, cellulose nitrate plasticised with camphor, a natural product, was a well-established material for a variety of applications. As long ago as 1901, the search for technically superior and price stable substitutes for camphor resulted in the first patents for phthalate[4] and phosphate[5] esters as plasticisers. During the 1920s, the mass production of cars led to increasing demand for plasticised nitrocellulose paints. By this time, the two most important industrial plasticisers were tricresyl phosphate (derived from coal tar phenols) and dibutyl phthalate. The possibilities for phthalates as industrial chemicals had been increased by the introduction in 1919 of air oxidation processes for producing phthalic anhydride.

   Polyvinyl chloride was first produced in the middle of the nineteenth century and was initially little more than a curiosity. The idea of plasticising PVC to give a useful material was patented in 1913, although twenty years elapsed before commercial exploitation occurred. A significant event occurring in 1929 was the patenting by Monsanto of di-2–ethylhexyl phthalate with plasticisation of nitrocellulose still being the primary aim. The scale of growth of the plasticiser, together with that of the PVC industry over the succeeding decades, would not have been anticipated.

1931 to 1965: Birth, growth and diversity of plasticised PVC
The main factor holding back the technical exploitation of PVC was its poor heat stability. At the temperatures required for thermoplastic processing, it rapidly developed dark colour at the same time as emitting corrosive hydrogen chloride. Both external and internal (using vinyl acetate as the comonomer with vinyl chloride) plasticisation were tried before 1930 as a means of reducing the processing temperature in order to mitigate the instability problem. When it was realised in the early 1930s that this defect could be overcome by simple incorporation of stabilising additives such as alkali metal soaps, the barrier to development was overcome.

In these early years, PVC was only used in its plasticised form and was regarded as a rubber substitute, particularly in cable insulation. Existing rubber processing machinery was modified where necessary to compound and fabricate plasticised PVC compositions. For their plasticisers, the early PVC formulators turned to available materials routinely used with nitrocellulose, in particular tricresyl phosphate and dibutyl phthalate which proved adequate for their purposes.

During the Second World War, two developments in PVC occurred which greatly accelerated its growth. One was the appearance of 'paste' grade polymers which could form liquid dispersions at high concentrations in plasticisers. This opened the way to a wide range of new processes and end uses. The second was the development in Germany of technology for extruding PVC pipe, thereby providing a substitute for traditional materials. From this beginning, rigid PVC outlets eventually overtook plasticised PVC during the 1960s in terms of the tonnage of polymer consumed.

The realisation that PVC had an unprecedented capability for accepting and retaining plasticiser led to rapid expansion in research into plasticisers for extending its useful range or increasing its competitiveness. By 1943, about 20,000 materials had been mentioned as plasticisers in the literature. Monsanto alone is reported to have investigated 3000 potential products between 1945 and 1953.[6] From these vast numbers a small proportion were commercialised, some being interchangeable, other products being recognised for their unique combination of attributes. By 1950, DOP had become the largest tonnage plasticiser and together with a small group of very similar phthalates, it has remained unchallenged ever since.

The large variety of plasticisers commercially available by the mid 1960s allowed the formulator to design different PVC compositions capable of serving at temperatures from below -40 degrees C to +105 degrees C, in contact with oil, resistant to burning or to equatorial weathering. They could be semi-rigid or rubbery, insulating or antistatic, solid or cellular, crystal clear or densely opaque. Plasticisers could be selected which were deemed by relevant authorities to be safe in food contact or medical applications. This enormous versatility and the low formulation costs achievable made plasticised PVC the most ubiquitous of plastics.

Whilst PVC has been the dominant outlet for plasticisers since the 1930s, uses in other polymers developed in parallel. The development of poly-vinyl acetate emulsions for adhesives and coatings depended on the use of plasticisers, mainly dibutyl phthalate. Polyvinyl butyral, from 1936 the main interlayer material for safety glass, requires the use of large quan-tities of speciality plasticisers. The processing and end properties of various synthetic rubbers are enhanced by the incorporation of phthalates or more specialised plasticisers and their consumption in chlorinated and nitrile rubbers is significant.

1965 to 1995: Consolidation, rationalisation and maturity
A 1965 technologist able to travel forward in time to the present day would find few changes on studying a list of commercially available plasticisers and the technical literature provided by suppliers. There would be some new supplier names and some unfamiliar product designations. In addition, there would be references to new end use specifications and a greatly increased emphasis on health, safety and environmental aspects. However, all the chemical types of plasticiser and their various performance characteristics would look comfortingly famil-iar. On enquiring into the latest consumption figures, the time traveller would probably be surprised by the increased dominance of just three phthalates (one of which would be somewhat unfamiliar) and the rela-tively minor position of higher performance plasticisers.

Whilst no significant new classes of plasticiser have emerged in the last thirty years, the 1960s did see moves to alternative feedstocks to improve performance, quality or competitiveness of existing types. A new genera-tion of chlorparaffins based on linear feedstocks was introduced which were superior in all respects to the earlier branched chlorparaffins and gained a significant share of the European plasticiser market. There was a move away from coal tar phenols to synthetic alkyl phenols as feedstocks for phosphates.

Alternative routes to linear alcohols (for phthalate manufacture) appeared in competition with natural product derivation or the type of ethylene-based process which had been used until then. This led to considerably increased use of linear phthalates in applications where their superior performance characteristics relative to the established branched phthalates were found to be of advantage. However, they have never gained the mainstream position achieved by linear phthalates in the USA following the introduction in 1970 of C_7–C_{11} mixed linear phthalate at a relatively small premium over commodities. In some of their applications, linear phthalates faced competition for a period from blends of commodity phthalates with esters of by-product linear dibasic acids, the so-called 'nylonates'.

The linear polyester plasticisers, a relatively recent innovation in 1965, were established in specialised market niches but did not have the sustained impact which might have been expected at the outset. Trimellitates were known as interesting plasticisers in the sixties but were not commercially available. Based on trimellitic anhydride from Amoco, the trimellitates have since become widely used. They have a distinct combination of characteristics, differentiating them from their closest competitors, polyesters and long chain phthalates.

The most significant changes in the plasticiser scene have occurred in the area of commodity phthalates as a result of suppliers becoming more vertically integrated. In the sixties, many plasticiser suppliers had wide product ranges which included phthalates based entirely on externally purchased raw materials. However, increasing competition from producers with in-house phthalic anhydride and alcohols gradually forced independent esterfiers to withdraw from phthalates and to concentrate on speciality products with higher margins. Nowadays, integration all the way from olefin via alcohol to phthalate is the norm. The most important example of this trend has resulted in a large share of the plasticiser market being captured by the 'iso-alcohol' phthalates which are dominated by a global supplier with a strategic feedstock interest.

In 1965, 'polygas' olefins produced from refinery mixtures of C_3 and C_4 olefins were sold as raw materials for iso-octanol and isodecanol, used to manufacture DIOP and DIDP respectively. Exxon Chemicals, the polygas olefin producer, is now the world's largest producer of plasticiser alcohols and plasticisers. One of the two main products of this route, DINP, competes closely with DOP. The other, DIDP, familiar for much longer, has some more distinct technical applications but is interchangeable with DOP in others. The same polygas olefin source has been exploited to produce a range of more specialised branched chain phthalates ranging from C_7 to C_{13}.

In addition to rationalisation amongst the suppliers, the market has seen Europeanisation of user demand both inside and outside the EEC.

Producers of compounds, cable, flooring and sheet have increasingly been absorbed by multi-national groups. This has encouraged rationalisation of formulations and increasing uniformity in the use of plasticisers. Expansion of EC membership has had similar effects. It is largely forgotten (except by ageing plasticiser technologists) that before entry into the EEC, the UK was most unusual in consuming very little DOP. There was no domestic production of 2–ethylhexanol and imported material carried a significant tariff. The market was served by a close technical equivalent, Alphanol 79, a multi-component C_7–C_9 alcohol mixture produced by ICI from wax-cracked olefins. The derived phthalate, known as DAP, was for some years the largest tonnage plasticiser used in the UK. It is now a fading memory, occasionally recalled because of residual confusion with diallyl phthalate, a small tonnage speciality monomer.

In 1980, the publication of work on DOP by the National Cancer Research Institute in the USA created widespread concern over a possible carcinogenic risk for humans. This stimulated large toxicological programmes on both sides of the Atlantic on DOP and other plasticisers. The eventual outcome in Europe was the official conclusion that DOP did not present such a risk and that restrictive classification should not be imposed. Details of these events are given in Chapter 9. However, before this conclusion became clear, there was a period of uncertainty resulting in some movement to alternative plasticisers free of the publicity focused on DOP (although not of proven lower risk).

A more systematic and sustained general influence has been the gradual introduction of legislation requiring suppliers to research fully the hazards of any new products and to fill progressively the gaps in knowledge of established chemicals. A milestone event was the establishment under EEC law of the European Inventory of Existing Commercial Chemical Substances (EINECS). This was taken as a definitive list of all chemical substances marketed in commercial quantities between 01.01.1971 and 18.09.1982. Materials missing inclusion on this list were then regarded as new and unknown. Any producer wishing to supply even small quantities of a new product for test marketing purposes was then faced with new costs in generating a prescribed set of HSE information. This has inevitably had some constraining effect on the appearance of new speciality plasticisers. However, the very maturity of the technology of external plasticisers means that the scope for innovation is in any case far less than it was forty years ago.

This section on recent history relates exclusively to PVC. A search of recent literature reveals many references to plasticisers in a much wider range of polymers, some of which were unknown thirty years ago. However, whilst plasticisers may make a vital contribution to the processing or end properties of these polymers, such uses are insignificant in plasticiser consumption.

1.6 MECHANISM OF PLASTICISATION

External plasticisers are not bound chemically to polymers and ideally are expected to be chemically inert. They can be recovered unchanged from compounds by extraction with suitable solvents. At the same time, the plasticised polymer appears to be an homogeneous solid with no sign of its liquid content and, if properly formulated, shows no change of physical properties over many years' service. With consistency of components and compounding conditions, the same physical properties are always achieved. In many instances consistency of properties is obtained even when processing conditions vary quite widely.

Different plasticisers display a very wide range of behaviour with the same polymer, both in the ease with which they form compounds and in the combination of physical properties which they confer. The two basic parameters of plasticiser interaction with any polymer are:

i) compatibility: to what extent is it capable of forming a stable compound with the polymer?

ii) efficiency: what degree of modification is achieved by incorporating a set level of plasticiser? Alternatively, what proportion of plasticiser is needed to achieve the required level of modification of a specific property?

Various overlapping theories were developed to account for the observed characteristics of the plasticisation process during the main growth period of plasticised PVC. These have been tested and refined right up to the present day. Any researcher or applied technologist working on plasticisers uses a combination of concepts from these different theories appropriate to the question addressed. Not surprisingly, most studies of the mechanism of plasticisation have concentrated on PVC although relevant part of the theories can be adapted readily to other polymers.

The transformation of solid particulate PVC plus liquid plasticiser into solid (apparently) homogeneous plasticised compound can be considered in the following stages:

- Physical mixing in which the plasticiser wets the particle surfaces but is not absorbed to any extent.
- Penetration of the particles which occurs as the temperature is raised.

The plasticiser becomes physically distributed throughout the PVC at a sub-particle level. At this stage, the composition becomes 'dry'.

• Separation of polymer chains occurs as the full processing temperature is approached. Dipole-dipole interaction between polymer chains is broken and plasticiser molecules form solvent type associations with the polymer. Some dissolution of polymer crystalline structure occurs. The composition is now in a fluid and formable state.

• Re-formation of polymer structure takes place on cooling of the processed compound to ambient temperature and a new structure containing the plasticiser forms. Eventually the structure reaches an equilibrium state (for practical purposes) which dictates the properties of the solid compound.

Plasticisation theories aim for understanding of the forces binding the polymer and plasticiser and the influence of the plasticiser structure on its compatibility and efficiency. A particularly wide review of the theoretical treatment of plasticisation has been presented by Sears and Darby.[7]

To be of practical value, a theory needs to account for differences in performance of different plasticiser types and to be capable of predicting the behaviour of plasticised compositions under real life conditions. With regard to the influence of plasticiser structure on performance, practising technologists tend to work with a very simplified set of ground rules. In their crudest form, they have been reduced to the notions that effective plasticiser molecules consist of 'sticky bits' (to bind them to the polymer) and 'floppy bits' (to separate the polymer chains and allow movement). To elaborate on this idea –

1. In order to be compatible with a polymer, a plasticiser needs to contain structural components which give loose reversible binding to the polymer, typically by dipole-dipole attraction

2. The remainder of the attached molecule, together with any unattached molecules, creates additional free volume in the structure which allows movement of the polymer chains and renders the material flexible.

Simple models indicating favoured polymer/plasticiser dipole alignments in plasticised PVC (as depicted two-dimensionally in Figure 1.1) were proposed by Leuchs.[8]

Much more recently, access to molecular modelling computer programs and solid state nuclear magnetic resonance spectroscopy has given indications of the more complex interactions which are likely to occur in reality.[9] Other studies in plasticised PVC at various stages of processing have shown that it is far from being an homogeneous material and that it possesses structure on a large enough scale to be detected by microscopy.

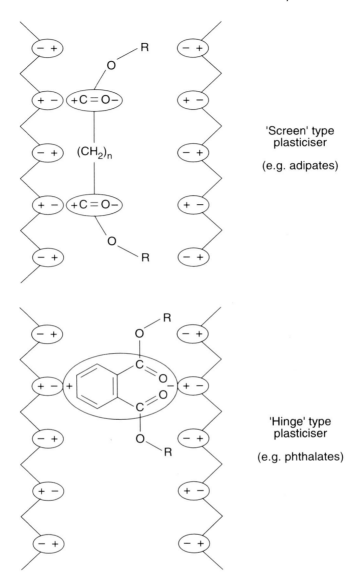

'Screen' type
plasticiser

(e.g. adipates)

'Hinge' type
plasticiser

(e.g. phthalates)

Fig. 1.1 Leuchs model of plasticisation.

Whilst such research continues to provide increasing insight into the structural complexity of plasticised polymers, there is still no universal set of rules to predict the performance of a potential plasticiser and there is still heavy reliance on analogy with known materials.

1.7   PREDICTION OF PLASTICISER/POLYMER COMPATIBILITY

In assessing whether a chemical of known structure is likely to be compatible with a polymer, the first step is often semi-intuitive, i.e. comparison with known plasticiser structures followed by a simple screening test. However, there are a number of parameters with a well founded theoretical base which are well recognised in plasticiser technology. These evolved from solvent theory and were based originally on the simple concept that 'like dissolves like'. In other words, for compatibility there must be a match between structural components in the plasticiser and the polymer. The following paragraphs give an extremely abbreviated treatment of this extensively studied subject.

Ap/Po RATIO: this is a simple quantitative extension of the above idea and was developed by Van Veersen and Meulenberg.[10] It is equal to the number of carbon atoms present in the plasticiser molecule (excluding aromatic carbon atoms) divided by the number of ester groups in the molecule. The reason for the exclusion of aromatic carbons from the calculation is the polarisable nature of aromatic groups which contributes to interaction. In this treatment, their effect is rated as neutral.

SOLUBILITY PARAMETER, δ this parameter can be calculated for a plasticiser or polymer by summating the effect of various structural components of the molecule. Thus:

$$\delta = \frac{\epsilon F}{V}$$

F is a constant devised by Small[11] for various common groups found in organic molecules and V is the molecular volume. As originally defined δ is directly related to cohesive energy density and different materials are mutually compatible when they have similar values for δ. Solubility parameters can be useful reference points for interpretation of experimental data. However in practice, deviations from predictions based on these values are usually found except when comparing plasticisers in a homologous series (for example the dialkyl phthalates). Differences in behaviour have been observed even between closely related triaryl phosphate isomers having the same calculated solubility parameter.[12]

Calculations of the above parameters take place without any reference to physical measurements. Some other means of prediction are based on physical property measurements of the plasticiser, for example dielectric constant.[13] Very many methods involving the use of polymer with plasticiser have been reported, for example measuring the equilibrium swelling of PVC compound immersed in plasticiser.[14]

The method most widely used to generate comparative data for commercial plasticisers is **solution temperature** (sometimes termed 'clear point'). This gives an initial indication of ease of incorporation of a plasticiser in PVC and the degree of compatibility in the compound. It is the temperature at which a mixture of plasticiser and suspension grade PVC appears to change to a single phase. The method has been standardised as DIN 53408 and commonly appears in plasticiser suppliers' literature. The significance of results from this type of test are discussed in Chapter 2.

The final practical check on compatibility is, of course, to attempt to make a compound and having done so, to observe it for separation of plasticiser. However, this moves ahead into the subject of **compatibility limits** which are discussed in Chapter 3.

1.8 POLYMER CHARACTERISTICS FAVOURING PLASTICISATION

With some types of polymer, notably polyethylenes, it is difficult or even impossible to produce plasticised compounds. In contrast, certain other polymers will accept unlimited amounts of plasticiser but increasing levels of addition destroy the ability to recover from mechanical deformation and the resulting soft sticky compounds are of little practical value. Polystyrene provides a good example of this behaviour. PVC is foremost among those polymers which are receptive to plasticisers but unlike polystyrene are still able to recover from various types of deformation even at very high levels of addition.

The resistance of polyethylene to plasticisation is characteristic of crystalline polymers. Polystyrene on the other hand shows behaviour typical of an amorphous polymer with nothing to constrain dissociation of the polymer chains after plasticisation. The value of plasticised PVC lies in the fact that it retains structural elements which constrain chain separation and thereby confer useful mechanical properties.

Simple gravimetric swelling measurements on plasticised PVC by Hecker and Perry[14] yielded a wealth of information, including clear indications of some form of crosslinking between polymer chains which was steadily reduced with increasing levels of compounded plasticiser. Immersion of a plasticised compound in plasticiser resulted in plasticiser absorption up to an equilibrium level. The higher the initial plasticiser content, the greater the subsequent uptake of plasticiser on immersion as shown in Figure 1.2.

The factors which constrain swelling (and also account for the tensile strength and recovery from deformation) of plasticised PVC are conven-

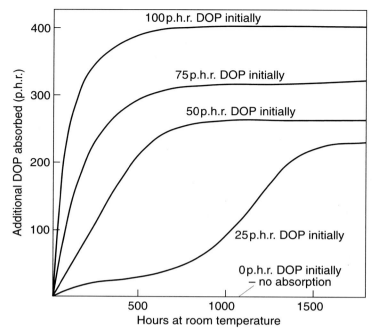

Fig. 1.2 Uptake of DOP by PVC containing various initial levels of plasticiser.

tionally attributed to small regions of order within the predominantly amorphous polymer. It is the amorphous regions which accept the plasticiser whilst the ordered 'crystallites' preserve the structure. Figure 1.3 gives a notional illustration of such structures in rigid PVC and plasticised PVC.

(In this way, plasticised PVC has a strong resemblance to the category of materials known as **thermoplastic elastomers** which were developed much later. These rely for their elastomeric properties on various types of non-covalent chain-chain association which is reversibly broken down by heating to processing temperature.)

The addition of low levels of plasticiser to a polymer can lead to an increase in crystallinity by lowering the energy barrier to rearrangement within amorphous regions to give ordered structures. This effect has been shown by X-ray diffraction to occur in PVC.[15] It accounts for the observed embrittlement when plasticisers are added at levels up to about 10%. The effect is termed **antiplasticisation**. Whilst it can be of benefit in optimising the mechanical properties of some engineering thermoplastics, anti-plasticisation of PVC is an effect which is normally avoided rather than sought.

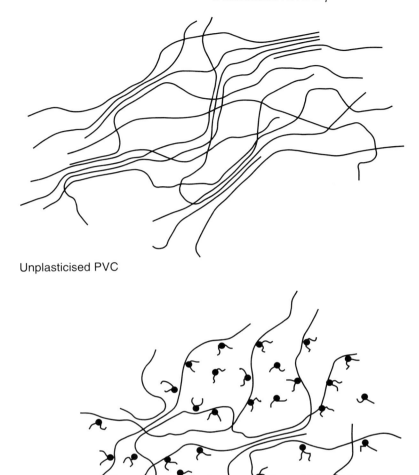

Unplasticised PVC

Plasticised PVC

Fig. 1.3 Notional structures of unplasticised and plasticised PVC showing retention of crystallites in plasticised polymer.

A notable exception to the earlier comment on the lack of utility of plasticised amorphous polymers is provided by polyvinyl butyral. As a result of its remarkable adhesion to glass, PVB is used as the standard interlayer material in car windscreens and architectural safety glass. The required energy absorbing characteristics are achieved by the use of high levels of external plasticiser. In plasticised rubber compounds, the constraints to chain separation are provided by permanent covalent cross-linking rather than the crystallites found in PVC.

Further reference to the mechanism of plasticisation is made at appropriate points in the following chapters.

1.9 THE FUTURE OF PLASTICISERS

So far we have defined the functions of plasticisers and reviewed the stages by which they have reached their current industrial status. We have also considered the mechanisms by which they are believed to achieve their useful effects. With this background Chapter 1 is concluded with a brief exploration of the factors driving and constraining future developments.

It is a safe prediction that external plasticising additives will be used in a variety of polymer for the indefinite future. However whether or not plasticisers retain their current high volume status is largely bound up with the future of flexible PVC. The influence of emotive attacks by pressure groups on PVC may now have passed its peak. These have been perpetuated by a belief that the production, use and disposal of the material was causing serious damage to the environment. In the future a cooler appraisal of the risks and benefits of PVC relative to alternative materials may hold greater sway. Nevertheless the negative publicity surrounding PVC has stimulated research into the use of substitutes in the main application areas of the plasticised polymer. Even with its negative environmental image set aside, such a mature product would be subject to gradual substitution as a result of normal technical evolution. It is generally considered that this process will be very slow.

Conversely the future competitiveness of plasticised PVC is strongly dependent on the continued availability of cost effective plasticisers which are not seen to pose unacceptable health, safety and environmental (HSE) risks. The market for PVC plasticisers has long been dominated by a small group of phthalate esters produced from C_8, C_9 and C_{10} alcohols. Between them these have given satisfactory performance in all the large tonnage applications of flexible PVC at prices for most of the time well below those of technical alternatives. Very few plasticisers match the balance of useful performance characteristics shown by the phthalates which, even with price parity might be expected to retain a major share of the market.

The general trend for end use specifications to become more severe will favour an increase in the use of C9 and C10 phthalates at the expense of the more volatile DOP.

As the major plasticiser class the phthalates (and in particular DOP) have been far more thoroughly researched for toxicological and ecotoxicological hazards than other types. Since there are obligations to make the findings public this has led to the perception in some quarters that the use of phthalates carries greater risks than the use of the less researched alternatives. Even in the absence of legal constraints on the use of phthalates this imbalance of information, which is likely to persist for a considerable time will be a factor favouring substitution provided the substitutes are economically competitive.

In view of the maturity of PVC plasticiser technology and the escalation in HSE research costs in commercialising new chemicals the appearance of unknown new types of PVC plasticiser in the future is unlikely. However the competitive position of known plasticiser types as reviewed in Chapter 5 could alter as a result of changes in feedstock prices, process technology and scale of production.

In some types of polymer the incorporation of plasticising comonomers offers a practical alternative to the use of external plasticisers if the use of the latter becomes constrained. The best current example of this is the use of vinyl acetate – ethylene copolymer emulsion adhesives in the place of plasticised PVAc homopolymers. Despite overcoming the intrinsic problems associated with traditional external plasticisers the disadvantage of this approach is a loss of formulating flexibility.

The unavoidable limitation of any externally plasticised composition is the potential for plasticiser loss under some service conditions. This not only alters the properties of the composition but, often more seriously can damage adjacent materials or have HSE implications. Whilst selection of plasticiser type can reduce losses to levels which are technically acceptable the solution is always a compromise. In general high permanence brings penalties of higher cost, reduced ease of processing and difficulties in achieving a satisfactory balance of physical properties.

One approach to avoiding the limitations caused by plasticiser migration would be the use of chemically reactive plasticisers. So called 'polymerisable plasticisers' have been in limited commercial use for many years. In their monomeric form they facilitate processing but once polymerised have little or no plasticising effect on the final product. It is possible that plasticiser technology could be influenced by developments in surface coatings which are being driven by restrictions in the use of volatile organic compounds. There is interest here in the use of reactive solvents which are incorporated into the structure of the cured coating. Some work on the grafting of plasticising side groups on to PVC polymer chains during processing has been reported in the literature.[16,17] A major challenge in developing this type of technology with PVC is in avoiding destabilisation of the polymer. An entirely different approach to preventing plasticiser migration would be surface treatment of the fabricated PVC product to give an impermeable integral skin without modifying its bulk properties (see Chapter 3, 3.8). An effective process of this type would allow the use of the most economical commodity plasticisers under a wider range of service conditions.

Any significant impact from the types of technology discussed above at present appears a remote possibility. The remainder of this book is focused on current processing practice and the wide range of established commercial plasticisers which continue to answer most formulators' requirements.

CHAPTER 2

Compounding and Physical Properties of Plasticised PVC

2.1 INTRODUCTION

There are two main process routes to plasticised PVC products, namely melt processing and plastisol processing. With the former, compounding of the various ingredients and the formation of the compound into its final shape take place at elevated temperature and require considerable mechanical work. In contrast, plastisols are liquid dispersions of the ingredients which are mixed and shaped at relatively low temperature after which the temperature is raised to effect the transition from the heterogenous polymer/plasticiser (etc) mixture to a homogenous compound. The criteria by which plasticisers are selected for optimum processing performance are somewhat different for the two routes. However, the mechanical properties of the end product show the same dependence on plasticiser type and level, regardless of whether melt or plastisol processing has been used.

For most applications, there is no choice of process route. For example, cable insulation is always made by melt extrusion and cushion vinyl flooring by plastisol spreading. However, for a few applications melt and plastisol processes compete. An example is PVC roofing membranes which are produced both by melt calendering/lamination and by plastisol coating. Most end use data provided by plasticiser suppliers are generated using test specimens produced by melt processing but are normally regarded as being equally applicable to plastisol-derived products.

The uncompounded polymer in the form of the white powder supplied is conventionally termed the 'resin' and consists chemically of the polymer plus low levels of characteristic impurities left over from the polymerisation process. Different grades of resin are designed for specific processes or groups of process or for specific end uses.

2.2 PVC RESIN PRODUCTION PROCESSES

2.2.1 SUSPENSION POLYMERISATION

The main process used to make PVC resins, accounting for more than 80% of production, is suspension polymerisation. In this process, an aqueous suspension of vinyl chloride monomer with droplet size in the range 50–150μm is formed by vigorous agitation in a pressurised vessel (required because of the low boiling point of the monomer (−13.9°C). The aqueous phase contains protective colloids to stabilise the suspension and buffers to control its pH. Polymerisation occurs under the influence of a monomer-soluble free radical initiator.

Particles of suspension PVC are roughly spherical with typical diameter in the range 50–250μm. They have a porous internal structure with a significant volume of interstitial voids capable of absorbing plasticisers. **Most resin used in melt processing of plasticised PVC is of the suspension type.**

2.2.2   BULK POLYMERISATION

This is sometimes termed 'mass polymerisation' and here the reaction medium comprises only the monomer plus dissolved free radical initiator. As polymerisation proceeds, insoluble polymer particles form and the monomer/polymer mixture is transferred to a second stage reactor designed for solids processing.

Whilst bulk polymerised resins have some intrinsic advantages over suspension types as a result of their higher purity and ease of plasticiser absorption, the process is difficult to operate and is not widely used.

2.2.3   EMULSION POLYMERISATION

Here the monomer is dispersed in an aqueous phase containing emulsifiers and an initiator. Initiation then occurs at the water/monomer interface. The polymer comprises small (<1μm diameter) spherical particles with a surface layer of emulsifier. The resin consists of aggregates of these primary particles which are formed during the spray drying stage of the process.

There is a variation on the emulsion process which employs a monomer-soluble initiator causing initiation to occur within the emulsified droplet. Although this is termed **microsuspension polymerisation**, the product has the general characteristics of a more normal emulsion polymer.

Emulsion resins have the characteristic of forming liquid **plastisols** (sometimes termed 'pastes') when mixed with plasticisers at ambient temperature. There is no absorption of plasticiser by the primary particles which retain their integrity until the temperature is raised to cause fusion. The difference between the particle forms of emulsion and suspension polymers is illustrated in Fig. 2.1.

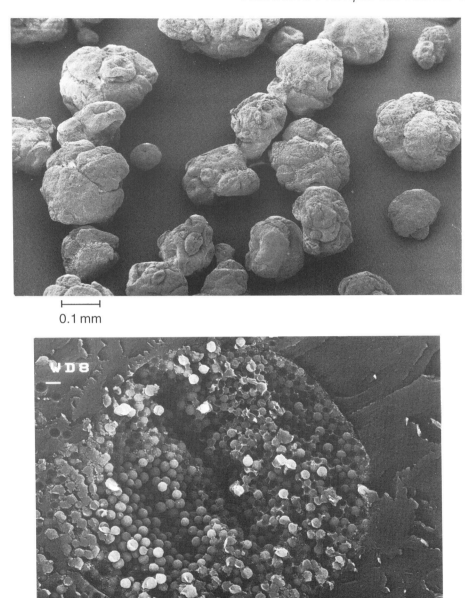

0.1 mm

10 μm

Fig. 2.1 Micrographs showing particle form of suspension PVC and emulsion PVC latex.
(by courtesy of European Vinyls Corporation (U.K.) Ltd.)

2.2.4 SOLUTION POLYMERISATION

Use of this process is confined to the production of relatively low tonnage speciality copolymers designed for use in surface coatings.

2.3 PVC RESIN VARIABLES

The important variables inherent in a PVC resin are its comonomer content, molecular weight, particle porosity and particle size distribution. Each of these has a bearing on plasticiser selection, either in processing or in the end properties of the composition.

2.3.1 COMONOMER CONTENT

The vast majority of PVC resin produced is homopolymer and there is strong reliance on additives, particularly plasticisers to exploit the wide range of properties achievable with the polymer. The most important comonomer used with vinyl chloride has always been vinyl acetate.

As mentioned in Chapter 1, the internal plasticisation conferred by vinyl acetate was originally seen as a means of reducing processing temperatures to avoid thermal degradation of PVC. The main practical use of VC/VAc copolymers was for many years the production of vinyl microgroove records on account of their excellent melt flow characteristics. Plastisol grades of VC/VAc copolymer resin have lower fusion temperature than homopolymers and are used in applications where process temperatures or times are limited. Examples are car underseal and coatings applied to temperature sensitive substrates.

Copolymer resins containing olefins are relatively unknown although they have been mentioned in recent patents.[18, 19]

Functional copolymers are sometimes used in blends with homopolymers to enhance adhesion to substrates or to participate in crosslinking reactions. Maleic acid and hydroxypropyl methacrylate are examples of functional monomers used for these purposes.

2.3.2 MOLECULAR WEIGHT

Standard indices of molecular weight are based on measurement of the viscosity of PVC solutions. The most commonly quoted are the ISO Viscosity Number and the Fikentscher K Value. Typical values are shown below, together with the corresponding approximate molecular weights for low, medium and high molecular weight resins.

The use of low molecular weight resins aids melt processing and is of

TABLE 2.1
INDICES OF PVC MOLECULAR WEIGHT

	low	medium	high
Viscosity Number	70	125	180+
K value	54	70	82+
Weight average molecular weight	70,000	200,000	500,000+
Number average molecular weight	40,000	64,000	90,000+

greatest value in unplasticised or very highly filled compositions. High molecular weight polymers are used when resistance of the end product to hot deformation is required.

2.3.3 POROSITY

The particle porosity of a PVC resin is critical in the dry blending stage of processing plasticised suspension or bulk resins, particularly with high plasticiser levels. It is often assessed directly by measuring the uptake of a standard plasticiser (DOP) at elevated temperature. This type of test has also been adapted to measurement of plasticiser performance using a fixed reference resin and is discussed later in this chapter.

2.3.4 PARTICLE SIZE DISTRIBUTION

Particle size and particle size distribution are the parameters which influence the rheological properties of plastisols, making them suitable for different types of process. For example, high speed application of thin coatings in wallpaper production imposes completely different requirements from dip coating of industrial gloves. It is common practice to use mixtures of resins to tailor rheological properties for some applications, particularly to obtain low viscosity in plastisols with high solid/liquid ratios. Here the use of a second resin of selected particle size, termed the 'extender' or 'blending' resin leads to tighter packing of particles within the surrounding liquid phase.

2.4 ADDITIVES (OTHER THAN PLASTICISERS) IN FLEXIBLE PVC

2.4.1 GENERAL

The main variables governing the physical properties of a flexible PVC composition are plasticiser type and level and these form the core subject of this book. In generating plasticiser performance data, suppliers generally confine themselves to simple PVC formulations in which only the plasticiser varies. However, in formulating for a specific set of end properties and process requirements, the user must take account of the many possible interactions between the various components of the formulation.

If an additive is miscible with plasticisers then it has the potential to contribute to plasticisation and allowance should be made for this in setting the plasticiser level. This is a characteristic of most organic additives. If, as is the case with inorganic materials, the additive is insoluble in plasticiser, then it will remain as a separate phase within the compound and will generally have the opposite effect (although this will be insignificant at low levels of addition).

A further potential effect of organic additives in plasticised compounds is an increased risk of incompatibility, i.e. separation of a mixture of additives (always containing plasticiser) from the compound surface layer (see Compatibility Limits, Chapter 3, 3.2).

This section reviews most of the categories of additives used in plasticised PVC and draws attention to potential interactions with plasticisers.

2.4.2 STABILISERS

Heat stabilisers are essential ingredients of any PVC formulation, rigid or flexible, melt or plastisol processed. They prevent thermal degradation (dehydrochlorination) of the polymer in processing and in service. Degradation of PVC proceeds by the loss of hydrogen and chlorine atoms from adjacent carbons in the polymer chain with the formation of hydrogen chloride. The resulting $C = C$ double bond weakens the adjacent C-Cl bond leading to the loss of further HCl and the formation of a second double bond. This sequence continues as subsequent C-Cl bonds are activated and a sequence of conjugated double bonds forms as shown in Fig. 2.2. It is these polyenes which are the coloured components of degraded PVC.

Heat stabilisers interrupt the reaction sequence by addition to the polymer chain to prevent propagation and in addition bind the released hydrogen chloride. Most stabiliser systems used in commercial PVC formulations contain two or more components which operate synergistically. The main groups of PVC stabiliser used in flexible PVC are, in order of decreasing volume (1993 situation) lead compounds, mixed metal compounds and organotins.

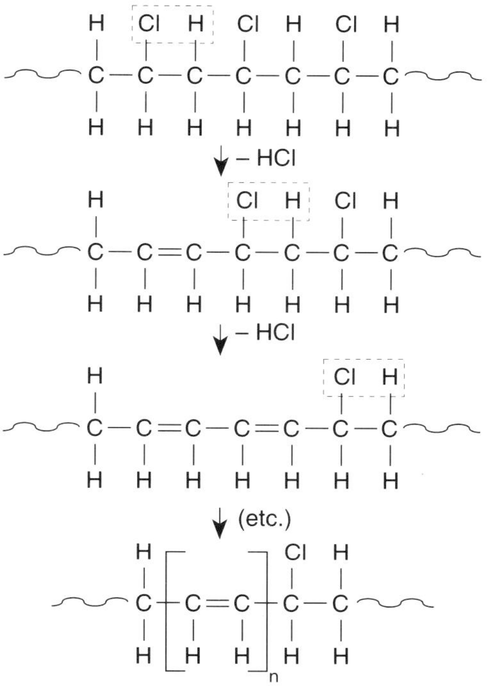

Fig. 2.2   Thermal dehydrochlorination of PVC.

2.4.2.1   *Lead compounds*

These are a range of basic lead salts of which the following are the dominant examples:

> basic lead carbonate
> tribasic lead sulphate
> dibasic lead phosphate
> dibasic lead phthalate

As a group, the lead compounds are the most cost effective PVC stabilisers, technically suitable for a wide range of applications and remain the normal type used in cable compounds. They are white solids and form a heterogeneous dispersion in the PVC. Their strong opacifying effect makes them unsuitable for use in transparent or translucent products. Lead stearates are dual function materials which act as lubricants as well as stabilisers.

In order to prevent hazardous airborne dust, lead stabilisers are often supplied in a damped or pasted form containing 5–20% of a general purpose plasticiser, such levels being too low to have any plasticising (or in rigid PVC antiplasticising) effect.

Lead stabilisers are gradually being supplanted by alternatives with lower levels of hazard (further reference to the concept of hazard is made in Chapter 9. Until now the pressures to do so have been less in Europe than they are in the USA.

2.4.2.2 *Mixed metal compounds*

These are soaps of certain metals used in specific combinations for synergistic effect. Formerly they were dominated by barium/cadmium and barium/calcium/zinc systems but the phasing out of cadmium use for HSE reasons has altered this picture. Important combinations currently used in flexible PVC include:

calcium/zinc – favoured whenever possible for low HSE risk
barium/zinc
barium/calcium/zinc

Recent developments in the use of co-stabilisers (alternatively termed secondary stabilisers or auxiliary stabilisers) have enabled suppliers to formulate mixed packages of sufficient efficiency for them to substitute cadmium- or lead-containing stabilisers.

The mixed metal soaps are solids, partially soluble in plasticisers, and they can have plasticisation and compatibility effects. Liquid grades are available in which the soaps are dissolved in liquid co-stabilisers (usually phosphite) and/or non-functional solvents.

2.4.2.3 *Organotins*

Most commercial organotin stabilisers are mixtures of mono- and di-organotin compounds, i.e. $RSnX3 + RSnX_2$. Here R is normally an alkyl group (e.g. n-octyl) and X is either a carboxyl group or a sulphur containing group (normally mercaptoacetate $-S-CO-CH_3$.

Organotin stabilisers are characterised by high efficiency (but high cost) and outstanding clarity in PVC compounds. Whilst they are widely used in rigid PVC, flexible applications tend to be limited by problems of odour (with the mercapto compounds) and compatibility (with carboxylates based on maleic acid).

2.4.2.4 *Auxiliary stabilisers*

There are additives which are of limited benefit when used alone but which have a co-stabilising function when used in conjunction with primary stabilisers, particularly the mixed metal type. The main such auxiliaries are:

Phosphites: $P(OR)_3$
The best known variant is triphenyl phosphite. Phosphites are believed

to function by chelating reaction products from the primary stabilisa-tion mechanism which would otherwise participate in degradation reactions.

Epoxides:
The main epoxy stabilisers are dual function additives which are also effective plasticisers. In the latter context, they are covered in more detail in Chapter 5. They are either epoxidised natural product tri-glycerides (e.g. epoxy soya oil) or esters derived from triglycerides (e.g. octyl expoxystearate). The reaction by which these additives stabilise PVC involves addition of hydrogen chloride to the epoxy group. The product of this reaction has low compatibility with PVC and can exude to give a sticky surface deposit.

2.4.3 LUBRICANTS

Lubricants are included in flexible PVC formulations, mainly to reduce adhesion between the hot PVC and machinery surfaces during melt processing and moulding. Their function here is described as **external lubrication** to distinguish it from **internal lubrication** of the composition. The latter improves processing by reducing inter-particle and inter-molecular friction. In flexible PVC the plasticiser serves this purpose.

Lubricants are materials having low solubility in PVC at melt process-ing temperatures. Hence they are capable of forming an adhesion-reducing layer between the melt and the metal surface at low levels of addition (< 1%). At the same time, there must be no sign, visible or otherwise, of lubricant incompatibility in the final product. This requires a fine balance of compatibility/incompatibility with the composition as a whole over specific temperature ranges.

External lubricants are waxy solids of the following types, having structures with low polarity:

hydrocarbon waxes
low molecular weight polyethylene
long chain alcohols (e.g. stearyl alcohol)
long chain fatty acids (e.g. stearic acid)
metal soaps (e.g. calcium stearate)
long chain amides (e.g. stearamide)
                                        etc

2.4.4 FILLERS

The fillers used in plasticised PVC are mainly cheap calcium carbonate types rather than the more expensive reinforcing fillers used in some other types of polymer. Their main function is usually cost reduction rather than beneficial modification of properties. A notable exception is the common use of calcined clays in cable compounds to increase their electrical insulating performance.

2.4.4.1 *Calcium carbonate fillers*

These are available in a wide variety of grades, produced by grinding and classifying chalks (whitings), limestone, calcite or dolomite. Their suitability for specific applications is governed by their chemical purity, crystallinity, maximum particle size, particle size distribution and the presence or absence of an applied surface treatment (normally stearic acid).

A key property quoted by suppliers is the capability of a filler to absorb plasticiser. This is normally measured as the amount of DOP required to wet out 100 g of filler to give a putty-like consistency (ASTM D 281–31). The use of any filler increases the hardness of flexible PVC and gives corresponding modification of other mechanical properties. The magnitude of the effect is related partly to the filler's plasticiser absorption and needs to be taken into account in setting the plasticiser level in a filled formulation. The effects of a coated ultrafine chalk filler on hardness and elongation at break of DOP-plasticised PVC are shown in Figs 2.3 and 2.4. It can be seen that increasing the plasticiser level by about 12 p.h.r. compensates for the effect on hardness of using 100 p.h.r. of the filler. However in the case of elongation no amount of additional plasticiser will compensate for the effect of high filler loadings.

A further important plasticiser/filler interaction effect occurs in plastisol processing. The inclusion of any filler increases plastisol viscosity but the effect is minimised by using low absorption fillers (coated and/or crystalline) with a particle size distribution conducive to close packing of solid particles in the surrounding liquid phase.

2.4.4.2 *Clays*

These are complex hydrated alumino-silicates containing minor impurities. The most common clay filler used with PVC is calcined kaolinite, the calcination process having dehydrated and partially purified the product. It is too expensive to use as a cost reducing filler and it has the disadvantage of contributing to extruder wear as a result of the particle hardness. However, its great merit is in acting as a scavenger for ionic impurities, thereby increasing the electrical volume resistivity of compounds for cable insulation. At around 10%, addition levels are much lower than for calcium carbonate fillers.

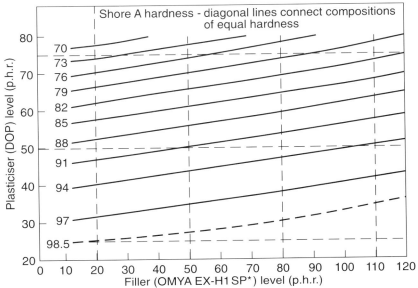

Fig. 2.3 Tensile strength contours for PVC compounds containing DOP and a coated ultrafine chalk filler.

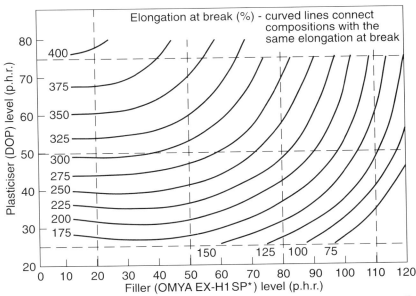

Fig. 2.4 Elongation at break contours for PVC compounds containing DOP and a coated ultrafine chalk filler.

* Croxton and Garry Ltd.

2.4.4.3 *Barium sulphate*

This is occasionally used as a filler for plasticised PVC in applications where its high density (4.5 compared with 2.7 for calcium carbonate) confers advantages of, for example, acoustic insulation or X-ray opacity.

2.4.5 COLOURANTS

Colourants are either dyes (soluble in PVC and dispersed at a molecular level) or, more commonly, pigments which are dispersed as fine insoluble particles. Soluble dyes have advantages in ease of dispersion and high clarity but give the risk of colour transfer by migration.

White and black pigmentation is provided in the main by titanium dioxide (in both its rutile and anatase crystalline forms) and carbon black respectively. Many other inorganic additives whose primary function is not pigmentation contribute to whiteness (e.g. lead stabilisers and fillers), thereby reducing the requirement for titanium dioxide.

For colours both inorganic and organic pigments are used. The inorganics have the better covering power (opacity) but lower colour strength. The use of some inorganics has declined at the expense of organics as users have dispensed with heavy metal compounds under HSE pressures. Because of their low light scattering, former organic pigments are often used in conjunction with titanium dioxide. Decreasing particle size increases both the colour strength and the transparency of organic pigments but also makes them more susceptible to migration. Some of these materials are slightly soluble in plasticisers and this can be sufficient to cause problems of blooming and colour transfer in plasticised PVC.

To facilitate handling, pigments for flexible PVC are often purchased as dispersed pastes in general purpose plasticisers.

2.4.6 FLAME RETARDANTS AND SMOKE SUPPRESSANTS

Plasticiser type and level are the most important variables influencing the fire resistance of flexible PVC and these factors are covered in Chapter 3.

The unplasticised polymer has high intrinsic resistance to burning as a result of its high chlorine content but plasticised PVC requires the use of flame retardant additives for certain applications. Phosphate esters and chlorinated paraffins are widely used as flame resistant plasticisers for PVC and the use of additional organic flame retardants in PVC would be unusual. However, inorganic flame retardants are used in flexible PVC, often in conjunction with these plasticisers. The most common of these additives are antimony trioxide and alumina trihydrate (ATH) which operate by different mechanisms.

Antimony trioxide acts synergistically with hydrogen chloride (available from the decomposition of PVC under fire conditions) to suppress gas phase combustion by free radical quenching.

ATH acts as an energy sink by dehydrating endothermically at high temperature. This action is augmented by the formation of a surface barrier layer of alumina. Compared with antimony trioxide, ATH has low efficiency and must be used at much higher levels, incurring the usual penalties of plasticiser absorption shown by fine particle fillers (see Fillers, 2.4.4).

Smoke suppressants operate by mechanisms resulting in the retention of carbon as a solid char at the expense of reactions (typically benzene generation) which lead to the formation of airborne particulates. As well as having a fire retardant action, ATH functions as a smoke suppressant at high addition levels as does magnesium carbonate.

The material most widely promoted as a high efficiency (although expensive) smoke suppressant for PVC, effective at low levels (ca. 5 p.h.r.) is **molybdenum trioxide**. **Iron oxide** pigments have a marked char forming and smoke suppressant effect which can make them cost effective for this purpose if their pigmentation and potential destabilisation effects can be tolerated.[20] **Ferrocene** derivatives which are known smoke suppressants for rigid PVC may function simply by acting as an easily dispersed non-pigmeting source of iron oxide. They are plasticiser soluble and generally too migratory and volatile for use in flexible PVC.

2.4.7 BLOWING AGENTS

These are additives which decompose during the processing of the composition with the evolution of gas bubbles, resulting in a cellular end product. They are incorporated into PVC plastisols used for the production of cushion vinyl flooring, foamed wallcovering etc. The blowing agents normally used in such applications are various grades of azodicarbonamide (AZDN) differentiated by particle size, the presence or absence of activators ('kickers') and their relative ease of dispersion. AZDN decomposes thermally between 200°C and 250°C to release nitrogen.

The function of activators is to lower the decomposition temperature to match the processing temperature range of PVC compositions. Many heat stabilisers used in PVC have this effect but for optimum activation, it is normal for the formulator to select an activator or to use a proprietary grade of AZDN containing one. The rate of nitrogen evolution is influenced by the activator and AZDN particle size (smaller = faster).

The main factor favouring efficient expansion and the achievement of uniform cell structure is a correct profile of melt viscosity of the composition versus temperature. This is strongly influenced by the degree to which the plasticiser solvates the polymer. Butyl benzyl phthalate (the most widely used fast fusing plasticiser) is conventionally used in cellular PVC products for this reason. The influence of the plasticiser here is physical and it has no chemical interaction with the blowing agent. However 'stabilised' grades of plasticisers containing hindered phenolic antioxidants can cause problems of colour formation as a result of reactions with AZDN and certain activators.[21]

2.4.8 ANTIOXIDANTS

The main group of polymers requiring stabilisation against oxidation is the polyolefins. Technical development of antioxidants which are effective at high processing temperatures and under end product service conditions is on-going.

An antioxidant functions by breaking the sequence of a free radical chain reaction involving C-H bonds and oxygen. In contrast to polyolefins, PVC does not require protection against oxidation. However, plasticisers can be susceptible to oxidation, initiated either thermally or photochemically. Free radicals generated in this process can in turn lead to dehydrochlorination of the polymer with consequent deterioration of physical properties. Hence it is normal practice to include an antioxidant in plasticised PVC formulated for service at higher temperatures, in particular in cable compounds.

The antioxidants used are of the hindered phenol type, the most common being diphenylol propane (DPP, Bisphenol A)

This material is widely available at relatively low price, being a major raw material for epoxy resins and polycarbonates. Because of its relatively low molecular weight, some volatile loss can occur at the highest service temperatures encountered but in practice this does not normally result in depletion to an ineffective level.

The alternative is to use one of the high molecular weight phenolic antioxidants developed for polyolefins. However, these are usually sev-

eral times more expensive than DPP and even at low levels of incorporation (ca. 0.1%) can impact on formulation costs.

In order to aid dispersion of antioxidant in the compound, it is common
for plasticisers to be supplied containing antioxidant (typically DPP in the
range 0.2–0.5%) if they have a tendency to oxidation (e.g. di-isodecyl
phthalate) or if they are intended for use at the high temperature end of
PVC's service range (e.g. tri octyl trimellitate).

There is an occasional misapprehension that the purpose of these
'stabilised' grades of plasticiser is improved storage stability. There is
rarely, if ever, a need for such stabilisation at ambient temperature.

In a flexible PVC formulation, additives other than antioxidant and
plasticiser can inhibit or promote oxidation. Carbon black, for instance, is
known to have an antioxidant effect by participation in free radical
reactions. Hence, any simplistic approach to formulating for oxidation
resistance can sometimes be misleading.

2.4.9 UV Absorbers

Photodegradation is a main element of the damaging effect of weathering
on plasticised PVC and is covered in the next chapter.

The first step in the process is the absorption by the composition of solar
radiation in the wavelength band 290–400 nm. The energy absorbed
results in bond fission and the generation of free radicals which then
participate in destructive chain reactions.

UV absorbers function by strong absorption of this radiation and
dissipation of its energy harmlessly as heat. In opaque compositions, the
pigments are effective to varying degrees in fulfilling this role, carbon
black being particularly efficient. However, in clear flexible PVC there is
potential for penetration by photons and initiation of chain reactions
within the bulk of the material and it is here that UV absorbers can be of
benefit. They are polymer soluble materials having high absorption
coefficients for wavelengths in the damaging UV range but are non-
absorbing in the visible range and are therefore colourless. The best
established UV absorbers are derivatives of either benzotriazole or benzo-
phenone. The absorption spectrum of one commercial additive of the
former type recommended for use in plasticised PVC is shown in Fig.
2.5.

There is ample evidence of the ability of these additives to suppress
colour formation and loss of mechanical properties in plasticised PVC.[22]
Some common plasticisers (e.g. benzyl phthalates, trimellitates) also
absorb strongly in the solar UV region and have been shown to have an
adverse effect on photostability. Since plasticiser concentrations in flexible

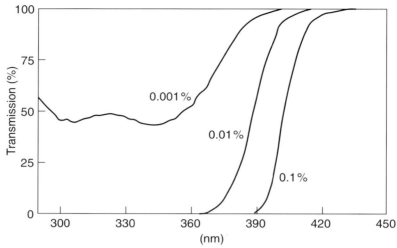

Fig. 2.5 Absorption spectrum of Tinuvin 571 (source: Ciba-Geigy Ltd.)

PVC are orders of magnitude higher than for a UV absorber, they might be expected to swamp its effect in competing for incident photons. Cases where UV absorbers are of benefit in the presence of UV absorbing plasticisers suggest that their stabilising effect is due not simply to the mechanism described above.

2.4.10 BIOSTABILISERS

The purpose of these additives is to prevent or inhibit the growth of micro-organisms on the surface of the PVC. The effects of unchecked activity range from conspicuous staining to significant changes in mechanical properties. In the absence of organic additives, PVC is resistant to microbiological attack. However, plasticisers and other additives with similar structural characteristics can act as a food source for micro-organisms. Flexible PVC may therefore need protection by the use of specialised additives in some circumstances. Conditions favouring microbiological growth are encountered in outdoor applications or generally in damp and unventilated environments. The mechanism of biodegradation is discussed in Chapter 3, 3.12.

The additives used to prevent attack are generally biocides. Ideally they should be specific in their effect, being very toxic to the target organisms but benign to all others. This is a very difficult balance to achieve in practice. The term 'biostabiliser' is preferred to 'biocide' since it indicates the intention to confine the effect of the additive to the plastic composition containing it.

Until the time of writing, the main biostabiliser for flexible PVC has been oxybisphenoxyarsine (OBPA), best known commercially in a diluted form as Vinyzene (Morton Thiokol). However, implementation of a European Community directive covering biocidal products is expected to exclude the use of arsenic-containing products in the near future and various options are being promoted as replacements.

A variety of materials constitute the active ingredients in these proprietary products, for example:

2–n-octyl-4–isothiazolin-3–one
(e.g. VINYZENE® IT – 3025 DIDP, Morton Thiokol)

2,3,5,6–tetrochloro-4 (methylsulphonyl) pyridine
(DENSIL® DOP, Zeneca Biocides)

zinc pyrithione
(zinc OMADINE®, Olin Corporation)

The first two examples quoted are supplied as dispersions in a general purpose phthalate plasticiser.

2.4.11   ANTISTATS

Flexible PVC is an excellent electrical insulator and this property is exploited in its largest application, low voltage cable insulation. Volume resistivity and surface resistivity values are dependent on the type and levels of additives present. Typically, surface resistivity values of flexible PVC compositions are sufficiently high to cause problems of frictional generation and accumulation of static charge which are characteristic of plastic materials in general. The charged surfaces prevent sheets from separating and attract dust. They also create a spark discharge hazard with consequent safety risks in the presence of flammable liquids and explosive gas mixtures.

The function of an antistatic agent is to dissipate the charge from highly charged areas by conduction across the surface or through the bulk of the material.

The most effective and permanent of antistats are certain grades of carbon black with low particle size and a high level of structure. Addition levels can be as high as 35% and plasticiser absorption effects need to be taken into account (see Fillers, 2.4.4). Carbon blacks are obviously only of use if a black product is required. Their level of performance cannot be matched in pale coloured compositions but many relevant antistatic specifications stipulating target values of surface resistivity can be met using organic additives. These antistats operate by two mechanisms:

(A) provision of a hydrophillic surface layer to facilitate conduction (this function is impaired by low ambient humidity);

(B) provision of ionic conducting species.

Two main chemical types of antistat are used in flexible PVC, cationics and nonionics. The cationics are normally quaternary ammonium compounds functioning by mechanism B. They operate most effectively in the presence of A-type antistats which are usually nonionic ethoxylated surfactants such as triethoxylated lauryl alcohol. Quaternary ammonium compounds have the drawback of promoting PVC dehydrochlorination and the resulting degradation colour precludes their use in clear and pale compounds. If an A-type antistat is used alone, it relies on traces of soluble ionic substances in other formulation components to provide the conducting species.

Organic antistats have limited PVC compatibility (indeed, this can be regarded as an essential attribute for the A-type) and excessive addition levels lead to surface exudation. Compatibility limits will depend on the plasticiser type and level and some trial and error is necessary to achieve an optimum balance (see Antistatic Plasticisers, Chapter 5, 5.2).

2.4.12 VISCOSITY REGULATORS

The rheological properties of a plastisol are governed mainly by the grade of resin and the plasticiser type and level. However, it may be necessary to use specific viscosity-regulating additives to aid mixing to match the viscosity to specific process requirements and to prevent or correct thickening during storage.

The use of volatile diluents in plastisols has been common in the past. These are materials of low viscosity which facilitate processing (most importantly, spreading) but evaporate during the fusion stage of the process.

In order to avoid bubble formation in the end product, their boiling point needs to be higher than the highest temperature reached by the coating. The main such materials, used on account of the low price, have been white spirit and similar hydrocarbon cuts. Dodecyl benzene (or linear alkyl benzenes) which is more expensive has been used for the same purpose, sometimes supplied as part of a proprietary plasticiser blend. It is less volatile than white spirit and a significant proportion is retained in the end product after processing. Because of its aromatic structure, it has high enough compatibility with PVC for it to be considered as a secondary plasticiser (see also TXIB, Chapter 5, 5.10).

HSE pressures have caused a decline in the use of these volatile (or semi-volatile) diluents both to reduce process emissions and to avoid exposure of the consumer to vapour emitted by the end product in service.

A different type of plastisol viscosity regulator reduces (and stabilises) viscosity by a surfactant mechanism. These additives are essentially the same type of material as the nonionic antistats, ethoxylated nonylphenol being a typical example. Addition levels are low (around 4 p.h.r.) but potential compatibility effects should still be considered.

The use of additives for the opposite effect, i.e. increasing viscosity or introducing non-dripping characteristics, is less common but can be useful in some dipping processes. The main examples of so-called thixotropic additives used for this purpose are fumed silica and aluminium stearate.

2.5   MELT PROCESSING OF PLASTICISED PVC

2.5.1   PROCESSES AND APPLICATIONS

Melt processing of PVC involves the formation of the molten composition into its final shape which is retained on cooling. Formation takes place by three main processes:

> extrusion
> calendering
> injection moulding.

The pre-requisite of any of these processes is the production of a uniform compound containing PVC and the various additives required to tailor performance to the demands of the end use and the formation process. The normal outline sequence of operations between raw materials and finished products is shown schematically in Fig. 2.6.

A very high proportion of flexible PVC compound used for extrusion and injection moulding is purchased by processors as ready compounded pellets. The largest suppliers of compounds are downstream businesses of PVC resin producers. There is a much smaller market for powder dryblend compounds for specialised processes, notably the slush mould-ing of automotive trim components which has grown rapidly in recent years.

Calendering is the continuous production of flat sheet by passing the melt through a series of (usually four) rolls. Embossing (to give surface texture) or lamination (to give a product of higher gauge or to apply to textile or other substrate) may be included in the process sequence either in-line or as a separate step. Major applications of calendered plasticised

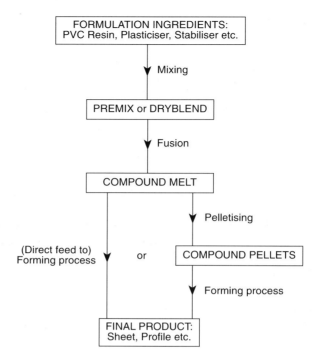

Fig. 2.6 Plasticised PVC processing sequence (melt processes).

PVC are floor covering, some types of automotive interior trim, roof sheet, liners for tunnels and reservoirs and stationery goods.

Extrusion is the production of continuous profile by forcing the melt through a die of the correct shape with a screw conveyor. Because of their lower melt viscosities, plasticised compositions are far easier to extrude than rigids and are normally processed using single screw extruders of relatively simple design.

By far the largest use of extrusion with a plasticised PVC is the application of insulation and sheathing in cable production. This requires the use of a crosshead arrangement in which the cable passes the tip of the extruder screw at a right angle. Other applications include various types of hose and tubing; refrigerator gaskets; water stop etc. An adaptation is the blow moulding of thin film for packaging and stretch wrapping in which a large diameter, thin-walled tube is continuously expanded by air pressure as it leaves the extruder dye.

Injection moulding forces the melt into a closed mould in the shape of the product. The process follows a cycle of cooling and opening for removal of the moulding. The main use of injection moulding of plasticised PVC is footwear production.

2.5.2 EFFECT OF PLASTICISER ON DRYBLENDING

If plasticisers and other liquid additives are mixed with an S-type (suspension) PVC resin at room temperature, the result is a damp powder which does not flow freely and tends to 'ball' when compressed, the more so at high plasticiser levels. In these mixtures, the majority of the liquid remains on or close to the surface of the particles. Provided that the process sequence does not necessitate the conveying of a free flowing powder, this 'premix' can be fluxed and further melt processed without problems. This is in fact normal procedure for small-scale laboratory compounding where the mix would normally be tipped on to a two roll mill for fluxing.

In order to produce a free flowing dryblend which can be handled easily, it is necessary to mix the ingredients at a temperature sufficiently high for the plasticiser and other liquid components to migrate from the external surfaces of the particles and become distributed throughout their interstices.

This can be accomplished using various designs of mixer with agitators having different peripheral speeds. The shortest mixing times and greatest efficiencies are found with high speed tank mixers where peripheral speeds up to $50 \, \text{m.s}^{-1}$ give a significant contribution to heating.

The ideal dryblend consists of resin particles into which the liquid ingredients have been completely absorbed, intimately mixed with particles of the other solid components. Thus any microsample should contain all the formulation ingredients in the correct proportions. It can be difficult to achieve complete absorption with high plasticiser levels and a technique sometimes employed in such cases is to withhold a portion of the resin until a late stage when it can 'mop up' the remaining surface plasticiser.

The ease of dryblend formation is critically dependent on the type of plasticiser in three ways. Firstly, the penetration and wetting of the inner surfaces of the particle will be influenced by the surface tension and viscosity of the plasticiser at the process temperature. Secondly, plasticisers of differing efficiencies are used at differing levels to achieve the required end properties. Third, and most important, are specific plasticiser/polymer interaction effects, most readily indicated by **solution temperature** which has been mentioned in chapter 1 in the context of compatibility.

The solution temperature is measured by a simple test which, in various forms, is widely used by plasticiser suppliers. It is always invoked at an early stage in the evaluation stage of any new material as a commercially useful plasticiser for PVC. The absolute figure obtained is of less significance than its position on an arbitrary scale relative to plasticisers of known performance. Such a ranking has a number of implications of practical significance. A low value of solution temperature indicates that:

- dryblends form at relatively low temperature;
- dryblending at a set temperature is relatively fast;
- powder mixes flux to homogeneous melts at relatively low temperature;
- optimum mechanical properties are achieved at relatively low temperature;
- plastisols (see section 2.6.5) fuse at relatively low temperatures/high rates;
- plastisols have poor storage stability, thickening with time;
- the plasticiser has high compatibility limit in PVC compositions and can be used at high addition levels without risk of exudation in service;
- the plasticiser is tolerant to the use of secondary plasticisers and other additives with limited PVC compatibility and gives low risk of exudation of such additives in service.

The basis of this test is observation of the behaviour of a mixture of S type PVC resin and plasticiser as the temperature is raised at a steady rate. The end point is indicated by the onset of a sudden change in physical appearance. The method has been standardised as DIN 53 408 and results based on this version are commonly quoted by German plasticiser suppliers. Non-standardised variables result in differences of up to 10°C in equivalent results from different users of this standard. This type of test has also been termed **clear point**, **apparent melting temperature** and **gelation temperature**. The variations involve a wide range of PVC/plasticiser ratios.

The method described by Anastagnopoulos[23] requires microscopic observation of a single particle suspended in plasticiser. The relevant temperature is that at which it melts and loses its identity in the surrounding medium.

A somewhat different test is one used by BP Chemicals in which a 25% dispersion of a medium K-value resin is agitated until rapid swelling of the particles and immobilisation of the suspension is observed. This occurs at a temperature typically 20°C lower than the solution temperature given by the DIN method. The gelation temperatures obtained by the BP Chemicals' method correspond closely to the swelling temperatures reported by Luther *et al* using a microscopic method.[24] Figures determined by different methods for a range of plasticisers are compared in Table 2.2.

TABLE 2.2
GELATION/SOLUTION/SWELLING TEMPERATURE
FOR VARIOUS TYPES OF PLASTICISER

PLASTICISER	Mol. Weight	Solution temperature DIN 53 408 (source: Huls)	Gelation temperature (source: BP Chemicals)	Swelling temperature (source: Luther et al)
n-butyl benzyl phthalate	312	–	88	87
di n-butyl phthalate	278	90	72	78
di-2-ethylhexyl phthalate	390	124	109	102
di–isononyl phthalate	419		118	
di–isodecyl phthalate	447	140	123	119
di-undecyl phthalate	478		136	
ditridecyl phthalate	530	160	146	147
di-2-ethylhexyl adipate	371	149	126	126
di–isodecyl adipate	427	164	143	
tri-2-ethylhexyl trimellitate	547		128	
linear C_{8-10} trimellitate	592	160	135	

These figures show a clear correlation with molecular weight for plasticisers within an homologous series, i.e. the gelation temperature of the phthalates falls as the active solvating part of the molecule is diluted with progressively longer non-solvating alkyl chains. The adipates and trimellitates show parallel trends.

The important conclusion from the foregoing is that solution temperatures etc. quoted by plasticiser suppliers are a valuable guide to dryblending and other aspects of PVC processing and compatibility provided comparisons are confined to figures from the same source.

Moving closer to real processing conditions, an instrument which is widely used to measure dryblending and rheological characteristics of plasticised PVC compositions is the Brabender Plasticorder. This is a torque rheometer with a variety of mixing head geometries. The adaptation used for quantifying dryblending properties is a planetary mixing head with the bowl surrounded by a heating jacket. Suspension resin is charged into the stirred bowl and heated to the test temperature before addition of plasticiser.

The most significant parameter is the time following plasticiser addition required to reach a minimum torque indicating completion of plasticiser absorption. This test as standardised in DIN 54802, ISO 4574 and ISO 4612 is designed to measure the capability of different resins to form dryblends with a standard plasticiser (DOP) at a fixed temperature.

Turning the test the other way round, i.e fixing resin and varying plasticiser, causes some difficulties in finding a set of test conditions to accommodate plasticisers with a wide range of solution temperatures. A temperature giving a clear result for fast solvating plasticisers can give an indistinct end point for plasticisers with much higher solution temperatures. Raising the test temperature to correct this compresses the scale for the more active plasticisers to an extent which makes comparisons difficult. However in practice it is unlikely that such widely differing plasticisers would need to be compared on a single scale.

Table 2.3 compares the dryblending times of a number of phthalate and trimellitate plasticisers under the following conditions:

Brabender Plasticorder planetary mixing head P600
Chart speed   1 cm/min
Mixer speed   100 r.p.m.
Temperature   115°C
Torque range   2 N.m

The instrument traces used to derive these results for two of the plasticisers are shown in Figs 2.7 and 2.8.

TABLE 2.3
PLASTICISER DRYBLENDING TIMES

PLASTICISER	Time (min)
Di-2-ethylhexyl phthalate (DOP)	1.6
Di-(linear C_7–C_{11}) phthalate	2.1
Di–isodecyl phthalate	2.9
Diundecyl phthalate	3.5
Tri-(linear C_7–C_9) trimellitate	3.1
Tri-2-ethylhexyl trimellitate	4.0
Tri-(linear C_8/C_{10}) trimellitate	6.0

2.5.3   EFFECT OF PLASTICISERS ON MELT FORMING

This stage covers the transition from powder blend to melt and the shaping of the melt into the final form. The term **'fusion'** describes the structural changes which take place within the composition during this sequence. Complete fusion is achieved when the original particulate nature of the polymer has reached a minimum and the dispersion of plasticiser at a molecular level is at a maximum. This condition leads to optimum tensile strength and extensibility in the product.

The process is complex and has been the subject of much research, some of which has focused on the influence of plasticiser type. General parameters by which melt processability is judged are the energy consumed in the process (related to the melt viscosity) and the visible uniformity of the

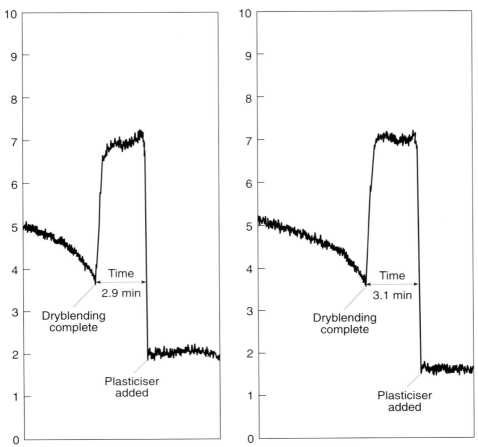

Fig. 2.7 Brabender plasticorder dryblending trace for di-isodecyl phthalate.

Fig. 2.8 Brabender plasticorder dryblending trace for tri-(linear C_7–C_9) trimellitate.

product (freedom from irregularities such as 'fisheyes', 'sharkskin' and the level of gloss). These effects were studied by Patel and Gilbert for extruded compositions containing seven different plasticisers at three different levels, 30, 50 and 70 p.h.r.[25] Their results showed the following:

a) power consumption (indicated by extruder motor current) decreased with increasing plasticiser level but had little dependence on plasticiser type;

b) bulk distortion of the extrudate, a defect associated with melt fracture,[26] was dependent on the PVC solvating characteristics of the plasticiser. The dependence varied according to the previous heat history of the composition.

c)   extrude gloss was achieved under conditions which produced good fusion as long as at least 50 p.h.r. of plasticiser was present. Butyl benzyl phthalate (known as the definitive fast fusing plasticiser for PVC) showed outstanding performance in this respect.

Because of the complexities of the transitions which occur, the interaction of resin, plasticisers and lubricants and the enormous range of machinery and process conditions available, there are no standard simple indices of the performance of plasticisers in melt forming processes.

It is important to ensure that the conditions of time, temperature and mixing efficiency are sufficient to achieve full fusion (or at least a degree of fusion sufficient to ensure specified properties for the end product). In general, this will be aided by the use of plasticisers with low solution temperatures but questions relating to surface quality etc. require more individual attention.

The great majority of general purpose and speciality plasticisers can be accommodated in melt processing sequences by readily achievable adjustments of processing conditions. In practice, plasticiser selection is dictated by end property requirements and differences in processing are not usually a relevant factor. In specific cases of difficulty, e.g. resulting from the high viscosity of polyester plasticisers, advice on optimisation of processing conditions is usually available from the supplier.

2.6   PROCESSING OF PVC PLASTISOLS

Plastisols are dispersions of the solid particulate components (predominantly PVC emulsion resins) of a formulation in the mixed liquid components (predominantly plasticisers).

They are themselves liquids with a wide range of viscosity, some being mobile and others having the consistency of non-drip gel paints. The parameters characterising this range of behaviour are set out in the following simplified treatment.

2.6.1   RHEOLOGICAL PROPERTIES OF PLASTISOLS

Consider a thin layer of plastisol of thickness d sandwiched between two parallel plates moving past each other with velocity V.

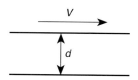

The plastisol is subjected to a shearing force which is opposed by its internal cohesion or viscosity. Assuming there is no turbulence in the plastisol, the resistance to shear will be proportional to V and inversely proportional to d. More accurately, it will be proportional to the differential dV/dx. This is termed the **shear rate, γ.**

$$\gamma = dV/dx$$

In practice, γ is approximately equal to V/x so that if a coating of thickness 0.5 mm were applied at a rate of 1 m.s^{-1} the shear rate would be

$$\frac{1 \text{ms}^{-1}}{0.5 \times 10^{-3} \text{m}} = 2000 \text{ s}^{-1}$$

The force per unit area required to producing uniform relative motion of the two plates is termed the **shear stress**, τ.

Viscosity, η, is the ratio of shear stress to shear rate.

$$\eta = \frac{\tau}{\gamma}$$

The SI unit of viscosity appropriate for plastisols is the Pascal.second (Pa.s). The traditional unit still widely used is the Poise (1 Pas = 10 Poise). Viscosities for plasticisers are much lower and are expressed in milli-Pascal.seconds (mPa.s) or traditionally in centiPoise, both units having equal value.

When this ratio is constant at all shear rates the liquid is termed **Newtonian**. Plasticisers are generally Newtonian but plastisols are not and as a result of their structure, show a range of deviation from this behaviour.

Dilatancy describes the increase of apparent viscosity with increasing shear rate.

The opposite effect, when apparent viscosity decreases as shear rate increases, is termed **pseudoplasticity**.

It is possible for complex liquids like plastisols to exhibit both dilatancy and pseudoplasticity over different shear rate ranges.

There are also time dependent effects where the apparent viscosity changes with time as a liquid is subjected to shearing at a constant shear rate. This is not to be confused with plastisol storage stability which is discussed later. Decrease of apparent viscosity with time is termed **thixotropy** whilst the opposite effect is **rheopexy**.

When a liquid will withstand a certain level of shear stress before deformation occurs, this threshold is known as the **yield point**. It is a

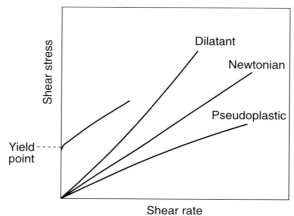

Fig. 2.9 Shear stress vs. shear rate for different types of liquids.

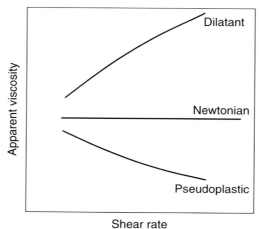

Fig. 2.10 Apparent viscosity vs. shear rate for different types of liquid.

necessary feature of plastisols for some applications such as non-drip or sag-resistant coatings.

These various types of rheological behaviour are depicted in Figs 2.9 to 2.11.

2.6.2 INDUSTRIAL PROCESSES

The common feature of plastisol processes is the fact that the plastisol is formed into the shape of the final product at temperatures close to ambient. After this, the temperature is raised to levels similar to those encountered in melt processing to effect fusion. Further mechanical

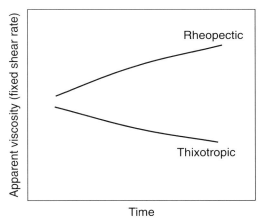

Fig. 2.11   Time-dependent viscosity.

deformation at the melt stage is relatively minor, e.g. roll embossing of a coating.

Because the mechanical stresses involved in plastisol processing are much lower than in melt processing, the machinery is less heavily engineered and capital costs tend to be lower. Plastisol processing also lends itself to a wider range of small scale operations.

The main large scale applications of plastisols involve flat coating of a wide variety of substrates (spreading processes).

Substrates	*End Use Examples*
paper	wallcoverings
glass fleece	cushion vinyl flooring
fabric: woven or non-woven, open or close weave, synthetic textile, glass, cotton	tarpaulins, upholstery, carpet backing
release surface	unsupported PVC for car interior trim
steel sheet (coil coating)	exterior building panels

There are various techniques for plastisol spreading grouped under the headings of knife coating, reverse roll coating and rotary screen coating. These are illustrated simply in Fig. 2.12.

Screen coating of plastisols expanded rapidly during the early nineteen eighties as a result of growth in the market for textured vinyl wallpaper. The technique forces plastisol from the inside of a thin walled hollow cylinder through finely spaced perforations on to the substrate. In

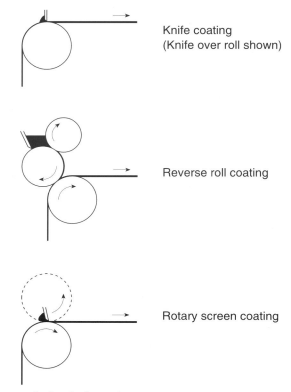

Fig. 2.12 Types of plastisol coating.

wallpaper printing a complex design can be applied by the use of a series of cylinders, each having a different pattern of perforations and applying a differently pigmented plastisol.

Coating processes typically involve high shear rates. For high speed application of thin coatings these can be well in excess of 10^4 s^{-1}. For such processes the rheology of a plastisol needs to be carefully tailored and this requires the use of special grades of resin. Once the coating has been applied fusion is accomplished by raising its temperature to (typically) around 180°C. This is normally brought about by passing through a hot air oven or less commonly by the use of infrared heaters.

The fusion stage of the process introduces two further important considerations for the plastisol formulator – firstly the rate at which full fusion can be achieved at different oven temperatures and secondly the level of volatile emissions from the process.

Thus there are four factors influencing the selection of plasticisers for plastisol processes, namely:

a) end property requirements;
b) plastisol rheology;
c) fusion characteristics;
d) process volatility.

As with melt processing, (a) is paramount but in plastisols the influence of plasticiser type on the process itself is clearer and more direct than in melt processes. Consequently it assumes greater importance. All four factors are discussed in more detail later.

After spread coating applications, by far the largest use for plastisols is automotive underbody sealants which are applied by airless spraying. This process requires low apparent viscosity at high shear rate together with yield point behaviour to prevent flow of the coating on non-horizontal surfaces before fusion. Again fusion rate/temperature and the level of process volatiles are critical factors which are dependent on plasticiser selection.

Other plastisol processes include the production of foam slabstock, dip coating (e.g. for making protective gloves), rotational moulding (for toys, footballs etc.) and a long tail of minor applications consuming relatively small quantities of material. Each of these has its individual set of process requirements under the four headings stated above.

2.6.3 MEASUREMENT OF PLASTISOL VISCOSITY

Because plastisols are not Newtonian liquids viscosity measurements should ideally be made at the shear rate corresponding to the process in question if they are to be usefully predictive. For some processes this is not practical since the shear rates are beyond the range achievable with available viscometers. However, comparisons obtained by extrapolation of apparent viscosities determined at lower shear rates give adequate guidance.

Viscometers range from simple quality control devices where the shear rate is not properly defined to sophisticated research instruments which not only measure viscosity over a very wide range of defined shear rates but also take account of their elastic behaviour.

The simplest of all measurements is by the use of a flow cup in which the time for a filled vessel to empty under gravity through a hole in its tapered conical base is recorded.

Much more versatile is the rotating spindle type of viscometer of which the Brookfield instrument is the most widely used. Here the resistance to rotation of a spindle immersed in the liquid is measured by a gauge calibrated to give a direct reading in Poise. Spindle speed can be varied to give an indication of shear rate sensitivity over a range of relatively low shear rates. Since the spindle rotates in a (theoretically) infinite volume of liquid with no opposing surface, simple values of shear rate cannot be calculated for such instruments.

The next stage of refinement overcoming this limitation is the bob and cup type of configuration in which the plastisol is sheared in the gap between a stationary cylindrical sample container and a concentric rotating cylindrical bob. The Haake Rotovisko using this type of system is used to measure plastisol viscosities at low to medium shear rates, i.e. up to about 10^3 s^{-1} but with lower limits for high viscosity plastisols.

Measurements at high shear rates can be carried out using cone and plate viscometers or gas extrusion rheometers. Cone and plate arrangements have intrinsic problems of sample ejection and high rotational speeds and with particulate liquids, such as plastisols, can give errors due to particle grinding in the narrow gap close to the cone apex. These can be avoided by the use of a truncated cone and by oscillating deformation involving small displacement of the sample. However, such refinements tend to be confined to expensive research instruments. The Severs gas extrusion rheometer which is widely used to measure plastisol viscosities at medium to high shear rates has the merits of being robust and relatively inexpensive.

Combination of results from this and from the Brookfield viscometer provide a profile of the rheology of a plastisol which can be related to most processes.

2.6.4   THE EFFECT OF PLASTICISERS ON PLASTISOL VISCOSITY

If there were no interaction between a plasticiser and suspended polymer particles, then the plastisol viscosity would be a function of plasticiser viscosity and would remain constant with time.

A close approach to this situation is found with weakly solvating plasticisers such as di-isodecyl adipate (DIDA). A strongly solvating plasticiser, dibutyl phthalate (DBP) which has very similar viscosity per se to DIDA (DBP 21 mPa.s at 20°C, DIDA 27 mPa.s at 20°C) shows completely contrasting behaviour. Figure 2.13 compares the Brookfield viscosity of plastisols comprising a microsuspension paste grade resin (Breon P.130/1) and 60 phr of plasticiser during seven days' storage at 23°C.

Of all the plasticisers used with PVC, DBP and DIDA represent the extremes of high and low interaction with the polymer. Whilst the low solution temperature of DBP indicates a variety of advantages in high compatibility plus rapid processing and/or low processing temperatures, the disadvantage of its high activity is its tendency to solvate plastisol resin particles even at ambient temperature. In Fig. 2.13, extrapolation to zero time shows that this process has already commenced during mixing. The viscosity increase is attributed to swelling of the resin particles, leading to

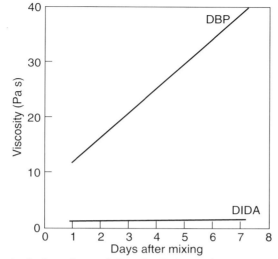

Fig. 2.13 Plastisol viscosity stability for DBP and DIDA.

decrease in the volume ratio of suspending liquid to suspended particles.

Resistance to viscosity increase is an essential attribute to plastisols which are stored for any length of time before use. It is particularly significant where plastisols are sold to processors with viscosity (or a viscosity-related parameter) included in the sales specification.

Highly solvating plasticisers would normally be avoided in such circumstances. One of the reasons why di-isodecyl phthalate (solution temperature 140°C) is preferred to di-2-ethylhexyl phthalate (solution temperature 124°C) for automotive underseal is superior viscosity stability.

The influence of DBP's solvating power on initial (before ageing) plastisol viscosity is an extreme case. For most plasticisers the more important factor is the viscosity of the plasticiser. The viscosity ranges of various groups of PVC plasticisers are shown below.

	Viscosity at 20°C mPas
General purpose phthalates C_8–C_{11}	80–130
Branched higher phthalates C_{11+}	200–300
Linear phthalates	40–60
Adipates	15–30
Trimellitates	140–300
Phosphates	20–200
Chlorparaffins	300–3000
Polyesters	2000–13000

It is possible to incorporate most types of plasticiser into plastisols but polyesters and the higher viscosity chlorparffins require dilution with materials of lower viscosity. At the other end of the scale adipates and similar aliphatic diester plasticisers are conventionally used by formulators as a means of achieving low plastisol viscosity.

The molecular structure of a plasticiser can influence plastisol viscosity in a way which is not simply related to thermodynamic compatibility with PVC. This point is illustrated by comparison of di-n-butyl phthalate (DBP), di-isobutyl phthalate (DIBP) and the plasticiser best known as Kodaflex TXIB which is 2,4,4-trimethylpentan-1,3-diol di-isobutyrate (see Chapter 5, 5.10).

DBP

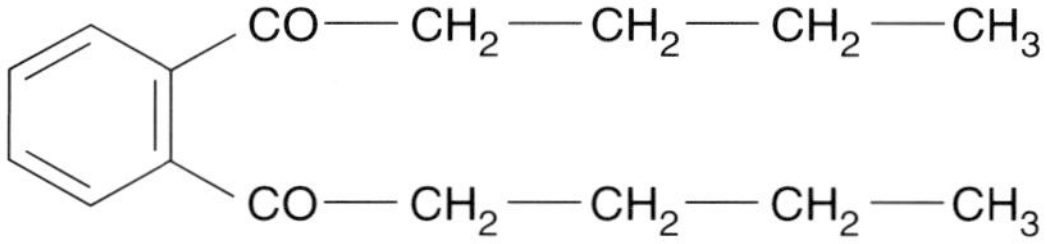

CO—CH$_2$—CH$_2$—CH$_2$—CH$_3$

CO—CH$_2$—CH$_2$—CH$_2$—CH$_3$

Molecular weight 278

DIBP

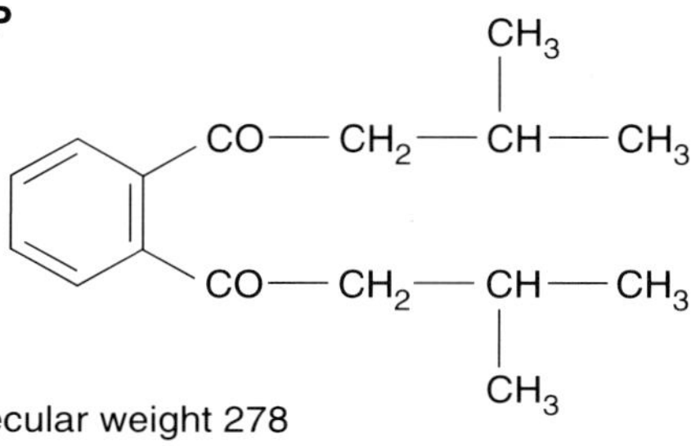

CO—CH$_2$—CH(CH$_3$)—CH$_3$

CO—CH$_2$—CH(CH$_3$)—CH$_3$

Molecular weight 278

TXIB

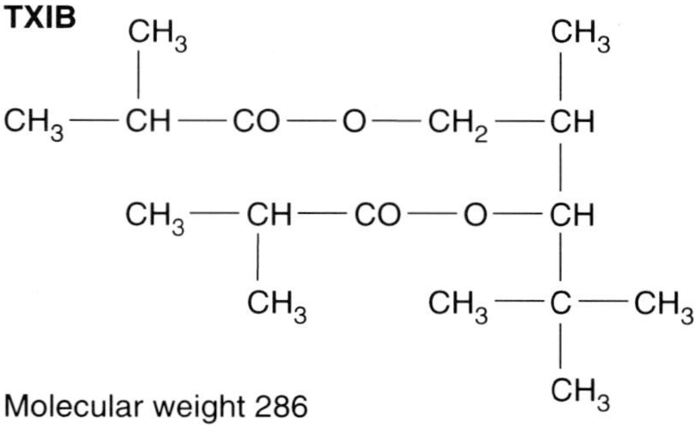

CH$_3$—CH(CH$_3$)—CO—O—CH$_2$—CH

CH$_3$—CH(CH$_3$)—CO—O—CH—C(CH$_3$)$_2$—CH$_3$

Molecular weight 286

DIBP differs from DBP structurally only in its alkyl chain branching. This gives it a significantly higher viscosity per se (39 mPas at 20°C versus

21 mPas at 20°C for DBP). However as illustrated in Fig. 2.14 it gives lower plastisol viscosity and greater viscosity stability than DBP and is therefore a better choice as a fast fusing plasticiser for plastisols. The difference in performance is probably due to kinetic factors.

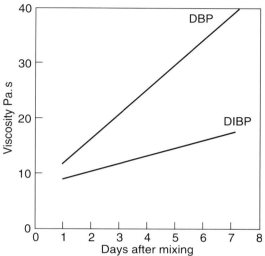

Fig. 2.14 Plastisol viscosity stability for DBP and DIBP (Breon *P* 130/1 100 parts, plasticiser 60 parts).

TXIB has very similar molecular weight to DBP and DIBP and like them contains two carboxylic ester groups. Its PVC interaction as predicted by its Ap/Po ratio of 7 (see Chapter 1, 1.7) would be lower than the two phthalates (Ap/Po ratio = 4) but should still be far higher than for e.g. DIDA (Ap/Po ratio = 12).

In practice TXIB is known for its outstandingly low plastisol viscosity and high viscosity stability and in this respect probably has the best performance of any commercially available plasticiser. These characteristics are undoubtedly related to steric factors. The high level of branching gives configurations which in the liquid shield the polar groups from interacting with the PVC but does not prevent solvation at processing temperatures.

Structural differences are a common factor influencing plastisol viscosity although their effect is seldom as marked as in the foregoing examples.

2.6.5   THE EFFECT OF PLASTICISERS ON PLASTISOL FUSION

As the temperature of a plastisol is raised progressively from ambient a point is reached when penetration of the dispersed PVC resin particles by plasticiser results in a degree of swelling sufficient to cause a noticeable increase in viscosity. Eventually the swollen particles will merge and at that point the plastisol will have 'gelled'. Thereafter the process of fusion commences. The plasticiser becomes dispersed throughout the PVC at a molecular level and particulate identity is lost. When the fully fused molten composition is cooled its structure eventually reaches an equilibrium state which results in optimum physical properties. These properties show the same dependence on plasticiser type and level as in a composition which has been melt processed.

At the lower temperature end of this process sequence the effect of rising temperature can be a fall in plastisol viscosity prior to gelation. Practical problems caused by this effect are dripping in non-horizontal coatings and strike through of coatings on open weave fabrics. Measurement of viscosity (or a viscosity-related parameter) for a temperature programmed sample of plastisol has been used to study this pre-fusion stage.[27,28] The results of such work add useful detail to simple predictions of relative performance from plasticiser solution temperatures.

Provided the plastisol contains only mutually soluble components then as it fuses it becomes transparent, the change occurring over a small temperature range. This effect has been used to study the affect of plasticisers on plastisol fusion both microscopically[29] and by the hot bench method used by Greenhoe.[30] With the latter a plastisol is laid on a hot bench having a temperature gradient. In both cases test data were correlated with physical properties of solid test specimens which had been fused under the same conditions. The results obtained by Greenhoe led to the important conclusions that:

- complete fusion is achieved only when the plastisol reaches a certain minimum temperature (which is plasticiser dependent),
- once this temperature is reached then fusion is complete and the fusion time is merely the time required to reach this temperature.

These conclusions more recently enabled Poppe[31] to make realistic estimates of the conditions required to achieve complete fusion of plastisols containing various general purpose plasticisers in industrial coating processes.

It is unusual for a single technique to give a picture of the behaviour of a plastisol over the complete range of gelation – fusion. The method developed by Hoy[32] observed the change in reflectance of a temperature-programmed sample as it passed from a glossy liquid, through a matt stage (gelation) and finally became glossy again as fusion occurred. Complete gelation/fusion curves have also been produced by the use of the Rheometrics Mechanical Spectrometer which can continuously meas-

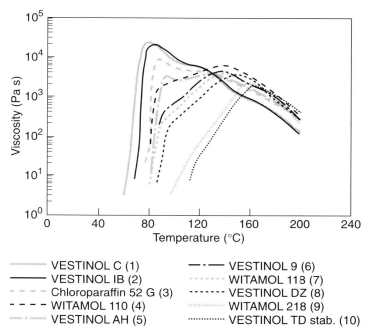

VESTINOL C (1)   — · — VESTINOL 9 (6)
VESTINOL IB (2)   WITAMOL 118 (7)
Chloroparaffin 52 G (3)   ------ VESTINOL DZ (8)
- - - - WITAMOL 110 (4)   WITAMOL 218 (9)
VESTINOL AH (5)   .... VESTINOL TD stab. (10)

Fig. 2.15   Effect of different plasticisers on gelation/fusion of plastisols.

ure the viscosity of a temperature-programmed sample subjected to oscillating deformation. Results published by Hüls AG using this method[33] give a particularly clear graphic comparison of the gelation–fusion behaviour of a wide range of plasticisers with a standard paste grade resin. These are reproduced in Figs 2.15 and 2.16.

These curves show a progressive change in position and shape with decreasing plasticiser activity. There is a remarkably good correlation with the solution temperatures determined by the simple DIN test using a suspension resin.

Technical literature from most plasticiser suppliers contains no fusion – related data beyond solution temperature (or similar test). However despite its simplicity and remoteness from real processes this test gives a very helpful indication of the order of merit of different plasticisers under a wide range of test conditions both for melt and plastisol processes. The actual figures are of less significance than the differences between them. Thus if a user changes from plasticiser A to plasticiser B with solution temperature 15°C higher, then processes will need to be adjusted to ensure that the new composition reaches temperatures about 15°C higher than the original in order to achieve the same degree of fusion regardless of the process details.

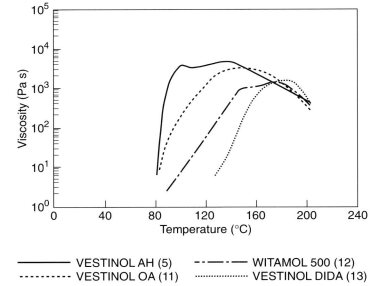

Fig. 2.16 Effect of different plasticisers on gelation/fusion of plastisols.

2.6.6 Volatile Loss of Plasticisers in Processing

The subject of plasticiser volatility is covered in more detail in the next chapter since resistance to volatile loss from the end product under service conditions is a key factor in plasticiser selection. In processing, temperatures are considerably higher than the highest temperatures encountered in service and rates of plasticiser loss are orders of magnitude higher. This can be seen from the vapour pressure/temperature plots in Fig. 2.17.

The consequences of high plasticiser volatility in processing are related mainly to health, safety and environmental (HSE) issues and in this context are discussed in Chapter 9. High volatile loss may result in a requirement for expensive containment measures to prevent exposure of operators to unsafe concentrations or the emission of unacceptable amounts of plasticisers to the environment. In extreme cases there is the simple economic question of waste if it becomes necessary to incorporate an excess level of plasticiser to compensate for process losses. In practice this last consideration is usually unimportant.

In any comparison of the processing losses of different plasticisers adjustments should be made for their different temperature requirements. For example at any fixed process temperature DOP is considerably more volatile than di-isodecyl phthalate but the difference can be reduced by the capability of DOP to achieve full fusion at lower temperature.

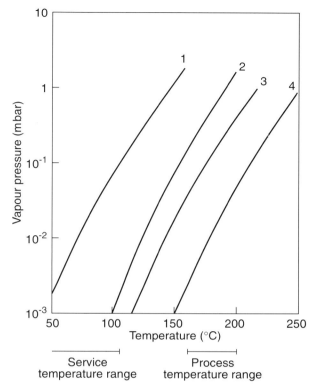

Fig. 2.17 Vapour pressure vs. temperature for representative plasticisers. 1 DPB; 2 DOP; 3 DIDP; 4 TOTM.

However Poppe[31] has shown by calculation that for the general purpose phthalates processing losses decrease with increasing molecular weight even allowing for such temperature adjustments.

There is a general rule (naturally with many exceptions) that easy processing and high volatility are linked. This forces one of the many compromises required in selecting the best plasticiser for a particular job.

2.7 THE EFFECT OF PLASTICISERS ON THE PHYSICAL PROPERTIES OF PVC

In all subsequent discussion of the effect of plasticisers on PVC we will continue with the normal technological convention of expressing plasticiser levels in parts by weight per hundred of resin (p.h.r.). This is more convenient than the use of % plasticiser/% PVC found in more scientific texts since it avoids confusion due to the presence of other additives. It also gives a more extended scale covering the range of practical interest.

2.7.1 MECHANICAL PROPERTIES

The useful range of plasticisation of PVC lies in the range of addition from about 20 to 100 p.h.r. With much more than 100 p.h.r. of plasticiser the material is limited by low tensile strength and is susceptible to soiling and surface damage unless heavily filled. Potential problems with processing and compatibility are further constraints on high levels of plasticiser. PVC compositions containing less than a threshold level of around 20 p.h.r. of plasticising additives are described as 'semi-rigid' and their uses are limited. With plasticiser contents between 0 and 10 p.h.r. compositions lie in the region of **antiplasticisation**. As discussed in Chapter 1 this is associated with maximisation of PVC crystallinity with about 4 p.h.r of plasticiser. The consequent effects of a typical plasticiser on mechanical properties are shown schematically in Fig. 2.18.

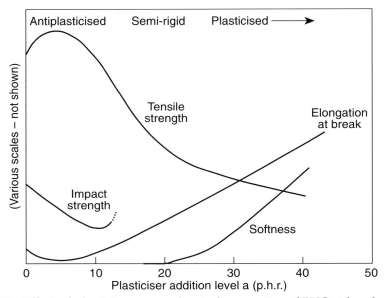

Fig. 2.18 Effect of plasticiser on mechanical properties of PVC at low levels of addition.

The physical properties of a PVC compound are dependent on temperature to an extent which is strongly influenced by plasticiser type and level. Thus full characterisation of plasticising performance would require tests for properties versus level at the full range of service temperatures which might be encountered in end use. Such data have been reported in the literature[34] but are not included in the routine data provided by plasticiser suppliers. This chapter concentrates on the interpretation of the type of routine comparative information available from trade sources.

2.7.1.1 *Hardness/softness*

The softness of a flexible PVC compound is a measure of the degree of penetration of its surface by a standard indentor under a dead load after a fixed time. Hardness is measured in the same way but uses a reversed scale giving high values for low levels of indentation. Softness measurements are standardised in BS 2782 and the commonly used Shore A Hardness test in ISO 868/DIN 53 505 and ASTM D2240. The relationship between the two scales is shown in Fig. 2.19.

Hardness is often the first criterion used to characterise flexible compounds. The efficiency of a plasticiser is then judged by the level of addition required for a specified value of hardness (or softness).

Softening efficiency shows a strong dependence on molecular weight and fundamental studies have related plasticising effect to molar concentration in the polymer. Polyesters do not fall within the same efficiency/molecular weight correlation as other plasticisers and are more efficient than would be predicted. Since plasticiser processing characteristics are also related to molecular weight then it is a useful rule of thumb that a plasticiser requiring high processing temperatures also has low plasticising efficiency. This then incurs the additional processing disadvantages which are a direct result of higher plasticiser levels.

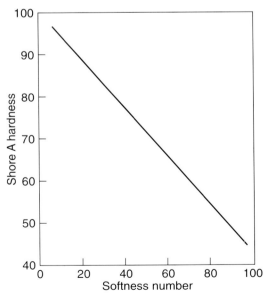

Fig. 2.19 Relationship between Shore A hardness and BS softness scales.

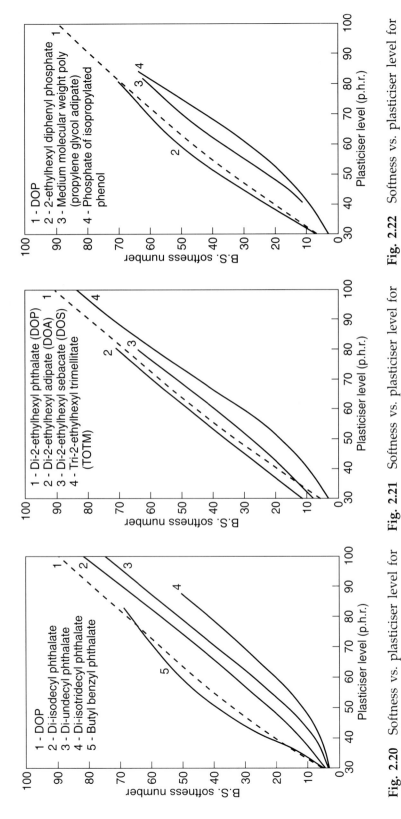

Fig. 2.20 Softness vs. plasticiser level for various phthalates.

Fig. 2.21 Softness vs. plasticiser level for various 2-ethylhexyl-esters.

Fig. 2.22 Softness vs. plasticiser level for various plasticisers.

The relationship between softness/hardness and addition level for a representative selection of plasticisers is shown in Figs 2.20–2.22.

Whilst these graphs show some interesting variations in shape due to molecular weight and structural differences, for most plasticisers they are roughly parallel straight lines over much of the composition range. This gives rise to another useful rule of thumb that increments of 1 p.h.r. increase softness number by 1 unit. This is helpful to the formulator when softness figures are only available for a single standard plasticiser level, typically 50 p.h.r or 66.7 p.h.r. (60% PVC/40% plasticiser).

2.7.1.2 *100% extension modulus*

This is sometimes used (although less commonly than hardness/softness) as the primary criterion for assessing a plasticiser's efficiency. Comparison of 100% modulus and hardness data published by a number of plasticiser suppliers for their product ranges shows little correlation between the two properties. Variation of 100% modulus (BS 6469, 1984) with plasticiser level for some common plasticisers is shown in Fig. 2.23.

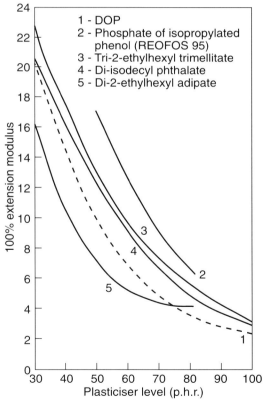

Fig. 2.23   100% extension modules vs. plasticiser level.

2.7.1.3 *Tensile strength and elongation at break*

Increasing plasticiser level reduces tensile strength and increases elongation at break as shown in Figs 2.24 and 2.25. Although these properties are affected to some extent by plasticiser type they are not normally factors dictating plasticiser selection. However measurement of tensile strength and elongation at break is a direct and practical way of determining whether an adequate level of fusion has been achieved in processing. Changes in these properties after heat ageing are of great importance in assessing the suitability of plasticisers for use at elevated temperature and this topic is covered in the next chapter.

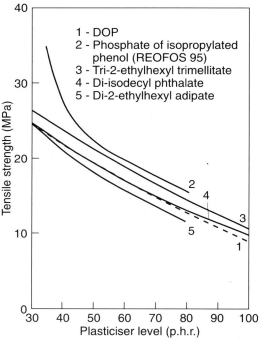

1 - DOP
2 - Phosphate of isopropylated phenol (REOFOS 95)
3 - Tri-2-ethylhexyl trimellitate
4 - Di-isodecyl phthalate
5 - Di-2-ethylhexyl adipate

Fig. 2.24 Tensile strength (BS 6469 [1984]) vs. plasticiser level.

2.7.1.4 *Low temperature properties*

With falling temperature plasticised PVC becomes progressively stiffer and over a small temperature range in the region of the glass transition temperature this change becomes steep. The mechanical response of plasticised PVC to temperature can be measured by determining torsional modulus over a range of low temperatures. The temperature at which it reaches a standard value is termed the **cold flex temperature**. The lower

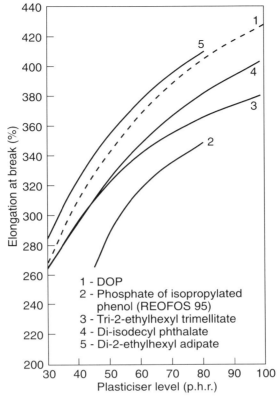

Fig. 2.25 Elongation at break (BS 6469 [1984]) vs. plasticiser level.

this value the better the resistance of the composition to stiffening under cold service conditions. In the standard test method, known as the Clash and Berg test after its inventors, the defined standard modulus is 310 N mm^{-2} and is indicated by a specimen being twisted through an angle of 200° under a standard load (ASTM D 1043, DIN 53 447).

The low temperature brittle behaviour of flexible PVC is determined by subjecting a sample of sheet bent into a loop to a standard hammer blow and examining it for fracture. This is done for samples conditioned at a range of low temperatures and the result recorded as the temperature below which fracture occurs (as in DIN 53 372). Various other types of low temperature test determine the temperature at which cracking occurs when specimens are subjected to some form of deformation relevant to the end use. The Clash and Berg cold flex test is the method most commonly quoted in plasticiser suppliers' literature although the user is often more interested in resistance to embrittlement. There can be a good correlation between results from the two test (see Fig. 2.26) although there are significant deviations in some published comparisons. In most cases cold flex temperatures tend to be 2–5°C optimistic as predictors of brittle temperature.

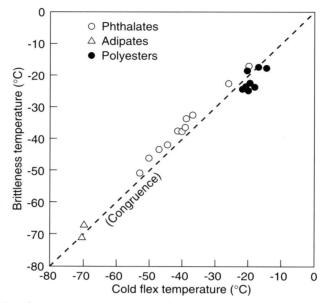

Fig. 2.26 Brittleness temperature (DIN 53372) vs. cold flex temperature (DIN 53447) for compounds containing 67 P.H.R. of plasticiser. Data Source: BASF AG[35]

Low temperature embrittlement sets a limit on the operating temperature range of flexible PVC products. Extension of this range is made possible by the use of plasticisers with good low temperature performance. This aspect of plasticiser behaviour depends on the molecular structure in a very clear way, being strongly related to linearity. The concept of linearity is discussed in detail for phthalates in Chapter 4, 4.8.1. There is no simple index of linearity which can be applied universally to all types of plasticiser but the following examples make the its influence clear qualitatively.

The effect of three types of plasticiser (see fig. 2.27a) on the low temperature flex point (cold flex temperature determined BP Chemicals modified Clash and Berg method) of PVC is shown in Figure 2.27b.

Plasticisers used as specialities for their low temperature performance are esters of the linear dibasic acids – adipic, sebacic etc. which have the general structure:

$$R–O–CO–(CH2)n–CO–O–R$$

HIGH LINEARITY dialkyl adipates (e.g. DOA)

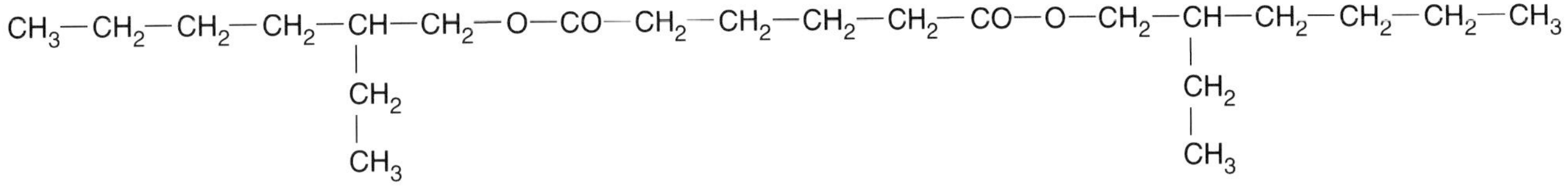

$CH_3-CH_2-CH_2-CH_2-CH-CH_2-O-CO-CH_2-CH_2-CH_2-CH_2-CO-O-CH_2-CH-CH_2-CH_2-CH_2-CH_3$

$\quad\quad\quad\quad\quad\quad\quad CH_2\quad\quad\quad\quad\quad\quad\quad\quad\quad\quad\quad\quad\quad\quad\quad\quad\quad\quad\quad\quad\quad\quad CH_2$

$\quad\quad\quad\quad\quad\quad\quad CH_3\quad\quad\quad\quad\quad\quad\quad\quad\quad\quad\quad\quad\quad\quad\quad\quad\quad\quad\quad\quad\quad\quad CH_3$

MEDIUM LINEARITY dialkyl phthalates (e.g. DOP)

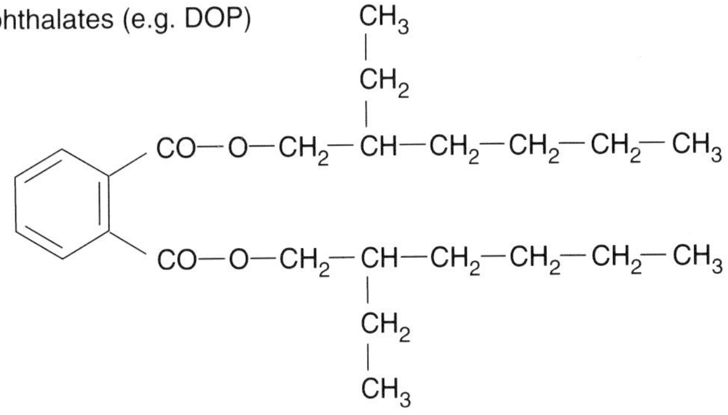

CH_3
CH_2
$CO-O-CH_2-CH-CH_2-CH_2-CH_2-CH_3$
$CO-O-CH_2-CH-CH_2-CH_2-CH_2-CH_3$
CH_2
CH_3

LOW LINEARITY triaryl phosphates (e.g. REOFOS 65® — typical component shown)

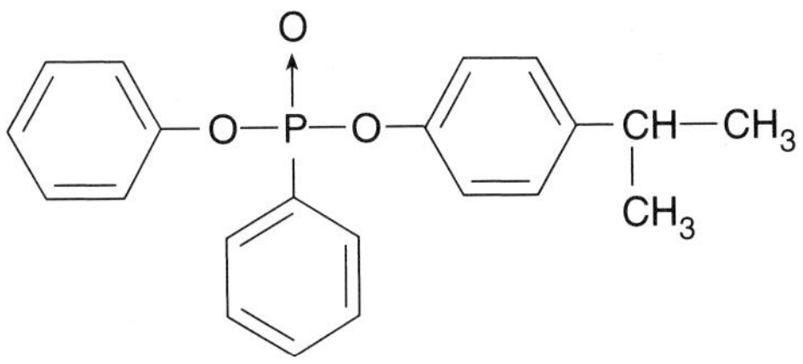

Fig. 2.27a   Structures of plasticisers having widely differing low temperature performance.

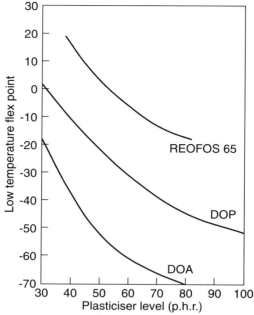

Fig. 2.27b Low temperature flex point (modified Clash and Berg) vs. plasticiser level for plasticisers representing extremes of linearity.

The ability of such materials to retain plasticising efficiency at low temperatures is closely related to their viscostatic character, i.e. their relatively small change of viscosity with temperature. The same factor is one reason for the use of di-2-ethylhexyl sebacate and esters of similar structure in lubricants designed to operate over a wide temperature range.

The use of low temperature plasticisers, generally in conjunction with other types of plasticiser allows the formulator to meet low temperature flex or brittleness specifications without using high plasticiser levels.

2.7.1.5 *Change of physical properties with time*

If softness measurements are made at ambient temperature on a specimen of flexible PVC at different times after moulding then a plot of the form shown in Fig. 2.28 is obtained.[36]

There is initially a rapid fall in softness which only after a considerable time appears to follow an asymptotic approach to a steady value. For this reason standards for measuring softness and hardness specify the conditioning time (7 days in the case of B.S. Softness). Long conditioning times give reproducible results but are clearly impractical for quality control purposes. On the other hand measurements made within a very

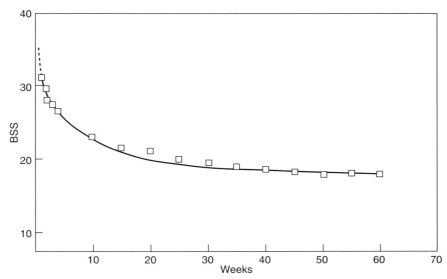

Fig. 2.28   Effect of aging time at room temperature on BS softness for PVC compound containing 50 parts of diisooctyl phthalate. Reprinted from Ref. 36, p. 118 by courtesy of Marcel Dekker Inc.

short time of specimen moulding give values on the steep part of the curve and can give rise to inaccurate predictions of eventual softness.

There is a corresponding change in all physical properties during this period although they generally reach steady values (within the limits of test accuracy) quicker than indentation softness. Conditioning times required for reproducibility are therefore shorter.

This hardening process is reversible and does not involve plasticiser loss. Reprocessing using the same conditions simply repeats the pattern. Providing an explanation for these observations has been an aim of the theories of plasticisation which are discussed in Chapter 1. Time dependent hardening is attributed to the development of a structure for the plasticised compound after cooling from the process temperature. In a constant environment the level of association between the polymer chains (crystallite content) and the distribution of plasticiser molecules in the PVC will progress towards an equilibrium state. For most practical purposes this can be regarded as complete within a few days although changes can be manifested after much longer times (see Chapter 3, 3.2).

2.7.1.6   *Plasticiser mixtures*
In this and succeeding chapters nearly all the figures showing performance of plasticisers have been obtained by tests on simple compositions containing a single plasticiser. The same is true of most reference data

provided in plasticiser suppliers' literature. However for many, if not most flexible PVC applications mixtures of two or more plasticisers are used in order to achieve the optimum combination of physical properties, permanence and processability at lowest cost.

Some data for plasticiser mixtures have been published, notably for chlorinated paraffins with phthalates.[37] In some cases plasticiser suppliers have offered customers formulating services using computer programs to identify lowest cost systems satisfying a specified set of requirements. However these tend to cover a limited range of options and the formulator usually has to rely on estimates based on standard data for individual plasticisers. It is normally a reasonable approximation to assume that no specific interactions occur and to use simple proportionation as shown in the following example.

ESTIMATION OF B.S. SOFTNESS AND COLD FLEX TEMPERATURE OF COMPOUND CONTAINING DOP 30 P.H.R. + BBP 15 P.H.R. + DIDA 15 P.H.R.

In this case the total plasticiser content is 60 p.h.r and therefore reference figures for this level of each component are required.

	Softness number at 60 p.h.r.	Cold flex temperature at 60 p.h.r.
DOP	46	−30
BBP	50	−13
DIDA	32	−58

$$\text{Softness number} = \frac{(30 \times 46) + (15 \times 50) + (15 \times 32)}{60} = 43.5$$

$$\text{Cold flex temp.} = \frac{(30 \times -30) + (15 \times -13) + (15 \times -58)}{60} = -33°C$$

Most reference data are obtained by tests on compounds containing no plasticising ingredients other than plasticiser. Allowance needs to be made for soluble organic additives. Since levels are usually relatively low it is usually a reasonable approximation to allocate them the same plasticising effect as (e.g.) DOP.

This simplistic type of calculation is not applicable to estimation of processing and permanence properties.

2.7.2 ELECTRICAL PROPERTIES

2.7.2.1 *Insulating properties*
Manufacturers of PVC covered cables are required to meet specifications guaranteeing the insulating performance of the covering. Such specifica-

tions may be stated in terms of tests which are not available to raw material suppliers. An example is the '**K Value**' which is related to the leakage of current from a 1 km length of cable immersed in water. However results of small scale tests measuring DC resistance are conventionally accepted as a reliable indication of performance. These tests measure electrical resistance between the faces of a flat moulded disc under an applied voltage of 500V DC (e.g. DIN 53 482 and BS 2782 202B).

$$\text{Volume resistivity (V.R.)} = \frac{\text{Resistance} \times \text{Area}}{\text{Thickness}}$$

V.R. is quoted in units of ohm.m or more often ohm.cm. The V.R. of flexible PVC is strongly influenced by both temperature and plasticiser content as shown in Figs 2.29 and 2.30.

Electrical conduction through flexible PVC depends on two factors, firstly the concentration of conducting ions in the composition and secondly the rate at which they are able to migrate through it. The second is related to the internal structure of the composition and is very dependent on plasticiser type. It is not generally realised that of the

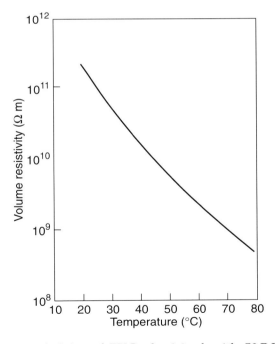

Fig. 2.29 Volume resistivity of PVC plasticised with 50 P.H.R. of DOP-variation with temperature.

a tenfold change in compound V.R. Within the normal range of industrial product the effect would be negligible. It appears that the main conducting species causing variation are likely to originate from compound components other than plasticiser.

2.7.2.2   Antistatic properties

The formulation of flexible PVC to give good antistatic properties has been discussed earlier in this chapter (2.4.11). In the quest for realism in testing some of the methods used for measuring antistatic performance can appear somewhat exotic. One such test used for antistatic flooring requires a participant to generate charge by repeatedly sliding about a large test area wearing large fluffy carpet slippers. However most specifications are more mundane and simply set a maximum limit for the DC surface resistivity, typically $10^8 \Omega$. This is determined by measuring the resistance between concentric electrodes of standard dimensions placed on one surface of a flat specimen (e.g. B.S. 3289). Surface resistivity depends on the same factors as V.R. but in addition it is greatly lowered by a high surface concentrations of components allowing conduction.

2.7.2.3   Dielectric properties

The high dielectric loss factor of plasticised PVC relative to polyethylene is a consideration which precludes its use in high voltage cable insulation or any other application where dielectric loss effects are unacceptable. Although there are differences between plasticisers in their effect on this property they are not normally large enough to influence plasticiser selection.

An area where high dielectric loss is an advantage is high frequency (radio frequency, RF) welding which is the process most commonly used for bonding plasticised PVC sheet. The amount of heat generated in a material subjected to an oscillating electrical field is related to the product of its dielectric constant and its loss factor has been termed the welding factor of a material. The effects of plasticiser on welding performance has been the subject of a number of studies which have been reviewed by Sears and Darby.[40] It is clear from these studies that the welding factor is not the only relevant variable determining the rate and ultimate strength of bonding. In practice compositions containing a variety of general purpose and speciality plasticisers can be welded successfully provided the process ensures even contact between the bonding surfaces and there is no migration of lubricating components to the surface.

An unusual application of RF heating of plasticised PVC is the highly specialised process for producing very thick section foam slab using volatile blowing agents. Since in the early stage energy is being absorbed by the liquid plastisol the 'welding factor' of the plasticiser per se is likely to be a relevant factor here.

2.8   SUMMARY – KEY PERFORMANCE INDICATORS

Ease of compound formation, processing characteristics and physical properties of a PVC composition are dependent to a large degree on the chemical structure and level of incorporation of the plasticiser. During the long history of plasticised PVC industrial and academic researchers have studied the effects of plasticisers on all manner of properties, some closely relating to major end uses, others more esoteric. The literature reporting this work can provide valuable guidance on many aspects of performance on which supplier information may be deficient. These sources have been thoroughly reviewed by Sears and Darby[41] among others. However even with access to such sources it is not always possible to find hard data for products of current interest relating to some aspects of performance. Throughout this chapter the emphasis has been on the standardised set of performance data provided by major plasticiser suppliers.

In practice there are strong inter-relationships between many perform-ance parameters and it is possible to provide a profile of the performance of an unknown plasticiser by testing for a very limited selection of key indicators.

2.8.1   MOLECULAR PARAMETERS

Before testing commences it is possible to make quite accurate prediction of performance by analogy with known plasticisers.

Molecular mass – Anything with molecular mass below 300 is likely to be too volatile for use in PVC and values above 800 (unless it is a polyester or other polymer) suggest low compatibility, difficult processing and low efficiency.

Polarity – Calculation of the Ap/Po ratio (Chapter 1, 1.7) will give an indication of processability and compatibility.

Linearity – If the structure is predominantly cyclic or branched then the material will show poor low temperature performance.

2.8.2   PLASTICISER PHYSICAL PROPERTIES

Measurement of viscosity at two temperatures provides valuable informa-tion on a plasticiser's performance. Clearly the results will give an immediate indication of its suitability for use in plastisols. At an early stage in the development of plasticiser technology the relationship between viscosity and plasticising performance was recognised by Leilich and became known as **Leilich's rule**.[42] In general low viscosity indicates high softening efficiency. Taking this a stage further a steep increase in viscosity with decreasing temperature suggests poor low temperature flex. In addition viscosity is a useful predictor of permanence, low viscosity indicating a tendency for rapid migration.

2.8.3   TESTS WITH PVC

2.8.3.1   *Solution temperature*
The relationship between solution temperature and a range of important parameters has been discussed (2.5.3).

2.8.3.2   *Softness/hardness*
If this is determined at a single level between 50 and 70 p.h.r. softness at other levels can be estimated by the use of the 'standard' softness/p.h.r. gradient. The softness of the compound will determine a number of secondary properties which are not normally measured such as its susceptibility to soiling and self-adhesion.

2.8.3.3   *Cold flex temperature*
Measurement at a standard addition level will give a ranking relative to known plasticisers. As well as pointing to good performance in all low temperature mechanical tests, a low value for cold flex temperature is indicative of low electrical resistivity and high general permeability.

2.8.3.4   *Volatility*
Measurement of volatility at a single addition level using the routine carbon absorption method (see Chapter 3, 3.4.3) clearly defines its suitability for various end uses and processes in relation to familiar standard plasticisers. Values significantly higher than for DOP may indicate limitations because of high process emissions or loss from the end product in service. In contrast a plasticiser with volatility lower than DIDP would be worth a more detailed evaluation for specialised high temperature applications.

   This brief reference to volatility is a convenient prelude to the next chapter which deals with the general subject of plasticiser permanence.

CHAPTER 3

Resistance of Plasticised PVC to Service Environment

3.1 INTRODUCTION

The last chapter dealt with the modifying effect of plasticisers on polymers. In this chapter we are concerned with plasticiser **permanence** and **stability**. The ideal plasticiser would show zero loss from the plasticised product and would remain chemically unchanged despite prolonged exposure to heat, light, aggressive chemicals, micro-organisms and contact with a wide variety of extracting media.

Any process causing plasticiser loss involves both thermodynamic and kinetic factors. Thermodynamics covers the strength of interaction of the plasticiser (its compatibility) with PVC relative to its compatibility with the medium into which it is migrating. In the case of volatile loss, vapour pressure is the measure of its 'compatibility' with air. Kinetics deals with the rate at which the plasticiser diffuses to the surface of the PVC and the rate at which it migrates away from the surface into the adjacent medium.

Loss will be increased by any factor which degrades plasticiser molecules into smaller fragments. These will have lower PVC compatibility, higher mobility and perhaps greater compatibility with extractants than the plasticiser itself.

The ultimate degradation of plasticisers occurs under fire conditions. Plasticiser type and level are the major variables governing the fire performance of PVC compositions. Although resistance of flexible PVC to burning is not strictly an aspect of plasticiser permanence the topic is included at the end of this chapter for convenience.

Before considering the many hostile environmental factors which can remove the plasticiser from the polymer it is necessary to return to the subject of compatibility. The first step is to ensure that the PVC composition has been formulated with regard to *compatibility limits*.

3.2 COMPATIBILITY LIMITS

The compatibility limit of a plasticiser can be defined as the level of incorporation in PVC above which it will exude or sweat from the surface of a fully processed compound. Exudation may first appear at any time after cooling of the processed article to ambient temperature. If the surface is wiped clean further exudation will continue to occur. In the early stages this is usually in the form of fine oily droplets but this can progress to form a continuous liquid film. No statement of compatibility limit can give anything better than a relative value since it will be dependent on the process, environment and above all formulation variables. For general purpose phthalates, phosphates and many other plasticisers compatibility limits are far in excess of the levels which would need to be used in practice. However for others exemplified by chlorinated paraffins and some adipates compatibility limits place a real constraint on their use. This distinction has led to the traditional classification of plasticisers as 'primary' and 'secondary'. A primary plasticiser is one which can be used on its own up to the highest addition levels ever required in PVC. In contrast a secondary plasticiser can normally only be used in conjunction with a primary plasticiser. Secondary plasticisers are used to contribute some specialised performance characteristic (e.g. good cold flex for DIDA). If they are used simply to reduce the overall cost of the plasticiser system they are termed 'extenders'. Chlorinated paraffins enhance fire resistance as well as reducing cost and so can carry either designation. Naturally the marketers of such products prefer the more distinguished term! In practice secondary plasticisers can often be used alone provided other formulation ingredients do not unduly depress their compatibility limits. For example chlorinated paraffins are used successfully as sole plasticiser in hard highly filled floor tiles.

The categorisation of a material as a primary plasticiser implies that its compatibility limit is not a matter of concern. However such plasticisers can show differences when used together with another material of lower compatibility. This characteristic is covered by the term 'extender tolerance'. Fig. 3.1 compares the compatibility limits of a number of phthalate plasticisers compounded in PVC together with Cereclor S52 (52% chlorinate of C_{14-17} n-paraffins). The plots were constructed by observation of a large number of moulded sheets containing the two plasticiser components in different proportions. Compositions below the curves are those which showed no exudation after six months.

Compatibility limits are elevated by the presence of fillers and to some extent by the use of lower molecular weight resins (Fig. 3.2). They are depressed by the presence of any soluble component having low interaction with the polymer, for example mixed metal soaps and epoxy co-stabilisers (Fig. 3.3).

For plasticisers having distinct melting points compatibility may be strongly affected by temperature. From consideration of its structure,

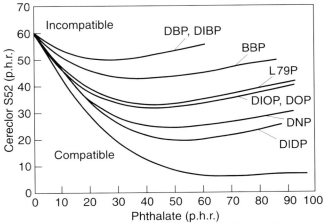

Fig. 3.1 Compatibility of Cereclor S52 in combination with various phthalate plasticisers. (Source: ICI Chemicals and Polymers Ltd.)

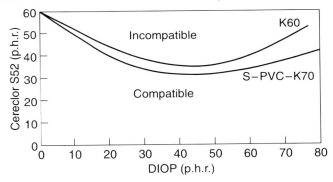

Fig. 3.2 Compatibility limits of Cereclor S52 with di-iso-octyl phthalate for polymers of K value 70 and 60. (Source: ICI Chemicals and Polymers Ltd.)

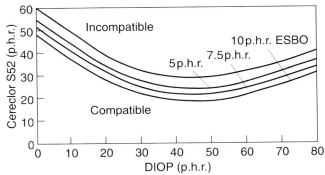

Fig. 3.3 Compatibility of Cereclor S52 with 2 p.h.r. barium/cadmium laurate and epoxy soya-bean oil. (Source: ICI Chemicals and Polymers Ltd.)

dicyclohexyl phthalate would be expected to have a very high level of compatibility with PVC. However it doesn't.

It is a crystalline solid with a melting point of 63°C. Whilst it can be compounded easily at high levels, when the compounds are stored at 23°C incompatibility appears in the form of a crystalline surface layer above about 30 p.h.r. This is indicative of the energy of crystallisation of the plasticiser being greater than its energy of interaction with PVC.

The same effect is a limiting factor in the use of long chain dialkyl phthalates. A phthalate made from 80% linear C12–C13 alcohols can be processed with PVC to give compounds with very good resistance to high temperature ageing combined with good flexibility at -40°C. At normal temperatures its compatibility limit is adequate enough to allow a useful range of addition levels. However, as a result of its high melting point (~15°C) conditioning of the compounds for any length of time at less than 0°C results in a waxy surface layer of the phthalate.

3.3 EFFECTS OF COMPRESSION AND MOISTURE ON COMPATIBILITY LIMITS

The previous section refers to the determination of compatibility limits for compound specimens which are in a mechanically unstressed state and free of environmental factors other than temperature change. However such factors need to be considered if exudation in service is to be avoided.

If a sheet of plasticised PVC is sandwiched between sheets of absorbent paper and subjected to pressure plasticiser will be lost to the paper. The rate of loss will be governed by the general factors discussed in the introduction to this chapter but will also show a clear dependence on the applied pressure. This behaviour is crudely analogous to water being squeezed from a sponge. In a practical sense the resistance to loss under pressure is a good indication of a plasticiser's compatibility.

It is the basis of the widely used loop compatibility test (e.g. ASTM D 3291) which simulates a condition known to have caused problems of exudation from finished goods in practice. In this test a specimen of sheet is bent into a tight loop and stored in this configuration for a set time. The inside of the loop is then under compression. The stress is not standardised here since it is dependent on the flexibility of the specimen which is of course dependent on the plasticiser. When the loop is opened the inside surface is examined immediately for exudation before the relaxation of stress results in reabsorption. One example of the use of this test has been its inclusion in specifications for compounds used in telecommunications wiring. In practice this is found to be very searching since of the plasticiser systems with acceptable cost performance very few are fully compatible under these conditions.

A less severe form of loop compatibility test has been used to determine guideline compatibility limits for formulators using chlorinated paraffins with primary plasticisers. Plots of the form shown in Fig. 3.1 are obtained but with the curves shifted to lower plasticiser levels.

Exudation of plasticiser can also be induced by high levels of humidity or immersion in water, conditions which are quite likely to occur in many applications. All but a few plasticisers used in PVC have extremely low solubility in water, typically less than 1 mg per g. Even with prolonged immersion, in the absence of surfactants, plasticiser loss is due less to dispersal in the water than to displacement from the PVC by absorbed water. The displaced plasticiser appears on the PVC surface as an oily exudate. Plasticised PVC is capable of absorbing a significant proportion of water depending on the plasticiser type and level. It appears that association between water molecules and the polymer can displace plasticiser. In general it is the least compatible (and most hydrophobic) plasticisers which are displaced to the greatest extent. However different influences come into play if the plasticiser contains hydroxyl groups or if it is susceptible to hydrolysis.

Plasticiser loss can also occur by displacement when PVC is exposed to vapours of other materials, for example hydrocarbons. Results obtained by a method developed to study the effects of different vapours have been reported by Darby *et al.*[43]

3.4 VOLATILE LOSS

3.4.1 VOLATILITY – GENERAL

Evaporation from PVC into the surrounding air is the most ubiquitous form of plasticiser loss. Since there is no chemical bonding between polymer and plasticiser evaporation occurs continuously even at low temperatures although it may be undetectable even by sensitive analytical techniques. The potential consequences of plasticiser volatility are two-fold. Firstly the plasticised product will become harder and less flexible to an extent which may be technically significant. In extreme cases there may be sufficient loss to result in noticeable shrinkage. Secondly the plasticiser vapour may have deleterious effects on adjacent materials or may pose a health, safety and environmental risk.

Standard tests for determining the effects of elevated temperature on flexible PVC involve measurement of weight loss or changes in physical properties or both. In compositions which are stable to oxidation plasticiser loss is the major factor causing physical property changes. The effect of plasticiser oxidation on physical properties is discussed later in this chapter. In any attempt to relate physical property changes to weight loss figures the latter need to be expressed as % loss rather than weight/area. Ignoring edge effects doubling of the thickness of a specimen of sheet will halve the reduction of plasticiser level although the loss per unit area will remain unchanged.

3.4.2 FACTORS AFFECTING VOLATILE LOSS

There are two major processes controlling the rate of volatile loss of a plasticiser from PVC, namely diffusion to the surface and evaporation from the surface. Under different conditions either one of these can be the rate determining step. For most applications in service evaporation is likely to be the controlling factor. Evaporation rate is dependent on the vapour pressure over the surface of the composition. Early studies showed that this had the same value as the vapour pressure over the uncompounded plasticiser, i.e. within certain limits it was independent of plasticiser concentration. Hence plasticiser vapour pressures should give a good indication of their rates of loss from PVC under service conditions. Table 3.1 shows the enormous range of vapour pressures found for plasticisers commonly used in PVC. The lower limit of measurement at 0.001 millibars represents for DOP a concentration of only $17.4 \times 10^{-3} \, \mathrm{mg} \, \mathrm{l}^{-1}$.

TABLE 3.1
VAPOUR PRESSURE OF PLASTICISERS AT VARIOUS TEMPERATURES
(Source: BP Chemicals)

PLASTICISER	Vapour Pressure (mbar) at:				
	80°C	100°C	140°C	180°C	250°C
DBP	0.02	0.082	0.87	6.08	86.67
DOP		0.001	0.039	0.571	20.75
DIDA		0.001	0.023	0.282	9.697
L79P			0.025	0.371	13.74
DIDP			0.009	0.143	5.144
L911P			0.005	0.087	4.099
TOTM				0.015	1.005

(DBP = di-n-butyl phthalate, DOP = di-2-ethylhexyl phthalate, DIDA = di-isodecyl adipate, L79P = di-(linear C7/C8C9) phthalate, DIDP = di-isodecyl phthalate, L911P = di-(linear C9/C10C11) phthalate, TOTM = tri-2-ethylhexyl trimellitate)

These figures provide a foundation for the well known rule of thumb that plasticiser loss doubles for every 10°C rise of temperature.

Diffusion becomes the rate determining step when conditions allow rapid removal of vapour from the surface. This will occur in a vacuum or under rapid air flow. The rate of diffusion is related to the plasticiser's molecular size and shape and to the permeability of the PVC. Permeability is strongly influenced by plasticiser level, diffusion being slower through a lightly plasticised composition than through the more open structure given by high plasticiser levels. This results in a progressive reduction in the rate of loss as the plasticiser content is deleted.

Evaporation controlled loss is linear with time and independent of plasticiser content. In contrast diffusion control gives linear loss versus the square root of time and shows a strong dependence on plasticiser level.

3.4.3 MEASUREMENT OF VOLATILE LOSS

In the development and use of tests for measuring plasticiser volatility there has always been something of a conflict between the needs for realism, reproducibility and convenience. If the air velocity over a sample of plasticised PVC is progressively increased then the rate of evaporation will increase to the point where diffusion control takes over and the rate of loss becomes independent of air flow rate. Royen[44] showed that with a particular design of tubular oven this occurred at about $1.5 \, \text{ms}^{-1}$.

Subsequently an oven operating with even faster air flow was developed by Pugh and Davis for testing compositions containing high temperature involatile plasticisers.[45] The merit of such designs is that they escape the effects of variables which can give poor reproducibility when evaporation control is acting. These problems are due to varying and undefined air flow over individual specimens and to a lesser extent to cross transfer of plasticiser between different specimens aged in the same oven. Whilst the ovens described above give excellent comparative data they give losses far higher than from other tests and lead to unduly pessimistic predictions of loss under real service conditions.

High flow tubular ovens are not generally available and are little used in standardised testing. In the USA box ovens with fast turbulent (as opposed to linear) air flow are available as standard items and are used for testing of PVC cable compounds against Underwriters Laboratory specifications. Like the fast flow tubular ovens they give relatively high volatility figures at any set temperature. They suffer the disadvantage of all box ovens, that is spatial non-uniformity of temperature and more importantly non-uniformity of air flow. In order to obtain reproducible results with a box oven it is necessary to confine specimens to a limited zone which can be located by testing identical specimens placed at different points in the oven. An alternative is to modify the oven by the addition of a turntable but this is not standard practice.

Specifications for heat ageing performance of plasticised PVC including maximum limits on volatile loss are most common in the cable industry. The influence of air flow is acknowledged by specifying limits on the number of air changes per hour in the test oven. However no account is taken of the oven size which is not specified. A significant example of this vagueness is found in a comparison of testing practice in Germany and the UK. As a result of international harmonisation of cable testing national standards in both countries require the same heat ageing performance when cable compounds are aged in an oven with between 8 and 20 air changes per hour. In Germany the oven which by custom and practice has become standard for this test is a box oven of internal volume 41 litres. Ovens are factory set by the manufacturer (Hereaus) to give the required air change by convection through inlet and outlet vents. In the UK the oven having parallel status (the Wallace oven) is of the multi-cell type in which each individual cell of volume 1.3 litres is swept at the required rate by a pumped air supply. Hence in the Hereaus oven specimens experience about 30 times greater air flow for the same nominal setting and it would be expected that volatile loss would be greater as a result. This is found to be the case as shown in the following example in which a cable insulating compound containing 60 p.h.r. of antioxidised DIDP was aged for 7 days at 100°C in both types of oven.

TABLE 3.2

	Hereaus Oven	Wallace Oven
Air changes per hour	8.3	20
Oven calibration method	power consumption meter	flow
Volatility (mg cm^{-1})	0.54	0.28

The main point to note from the foregoing is the need for caution in comparing volatility results from different sources even when the same test standard has been used.

Sensitivity to air flow, convection or forced, is avoided in the type of test which uses activated carbon to remove evaporating plasticiser. Here the test specimen is placed in close proximity to the carbon in a closed container at the set temperature. The widely used ASTM version of the test specifies 70°C for 24 hours. However higher temperatures and/or longer times are often used to give a wider scale of comparison. For the purposes of plasticiser evaluation carbon volatility is best regarded as an initial screening method for allocating plasticisers into categories of use. It would normally be followed by more searching tests if volatility were a critical factor in the envisaged application.

The volatility figures in Appendix 1 were produced by the activated carbon method described in BS 2782. Of the plasticisers listed DBP and DIBP are in a class of their own being regarded as too volatile for most PVC applications (and processes). Their volatility is more than 10 times higher than the general purpose phthalate plasticisers DOP, DINP and DIDP. The scale of comparison loses sensitivity for the least volatile plasticisers in common use, the trimellitates. The figures given by this test for polyester plasticisers can be somewhat misleading since they represent early loss of their volatile fractions. An interesting feature of the results is the general lack of variation of volatility with plasticiser level. This is consistent with the loss being under the control of evaporation as discussed earlier.

3.4.4   WINDSCREEN FOGGING

The topic of windscreen fogging has for many years found a place in any technical review or update of the plasticiser scene. It is well known that evaporation of plasticisers from car interior components can contribute to the phenomenon. However it is only one of a number of potential causes and the link between fogging and plasticisers can sometimes assume an exaggerated importance. Windscreen fogging is the obscuration of visibility through a car windscreen as a result of light scattering caused by deposits of airborne material on the interior surface. These deposits consist of condensate of organic vapours present in the passenger compartment together with adhering particulate matter.

Any material used for automotive interior component which contains any additive or ingredient with a discernable vapour pressure is a potential source of windscreen fogging. Some components may spend prolonged periods at elevated temperature as a result of solar heating or the car's air heating system. The most extreme conditions are encountered by the crashpad skin which is located directly under the windscreen. Here temperatures in excess of 100°C have been recorded.[46] Car manufacturers have developed a variety of tests to measure the propensity of interior component materials to cause fogging. The most commonly used form of test has been standardised as DIN 75 201. (At the time of writing this is also a draft ISO standard, ISO/TC45/13N-143). This test aims to simulate the conditions causing fogging by heating a specimen and collecting the emitted vapour as condensate on a cooled plate as shown in Fig. 3.4.

The standard gives two alternative methods of measuring the level of fogging. Method A determines the 60° reflectance of a glass collector plate

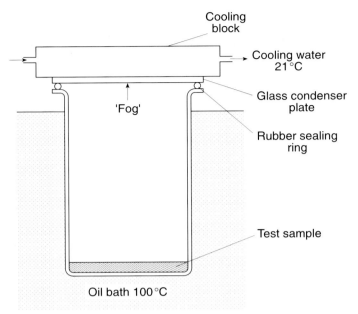

Fig. 3.4   DIN windscreen fogging test apparatus.

before and after fogging and expresses the result as a 'fog number' which is the % retention of reflectance, i.e. 100% indicates no fogging. The standard used to calibrate the test is a closely specified grade of DIDP. In Method B the collecting surface is a sheet of aluminium foil and the duration of fogging is extended from 3 to 16 hours. The result is expressed as the weight gain by the foil and the calibration standard is commercial DOP of typical high purity.

Whilst the reflectance method is more closely related to the actual fogging effect than the gravimetric method it is more difficult to operate and requires considerable operator experience. Reproducibility of the light scattering effect requires the deposition of condensate as uniform droplets having a clear contact angle with the glass surface. It has been pointed out by Jackowski and Poppe[47] that the surface tension of glass does not favour the formation of stable droplets. Some recent work has shown that reproducibility can be improved considerably by the application of a silicone coating to the plate.[48]

Windscreen fogging tests are designed for testing samples from finished trim components to determine whether they conform to limits set by car manufacturers. However the same test are used by producers of PVC trim for approval and quality control of raw materials. Low fogging by individual ingredients does not automatically guarantee low fogging by the finished product since the possibility exists of reactions occurring during processing generating sufficient volatile material to affect the result. For example any hydrolysis of a phthalate plasticiser would produce a relatively volatile alcohol (although in practice this would be most unusual). Conversely fogging by the final product would be better than anticipated if, as is often the case there were significant loss of volatile impurities during processing.

In the context of selecting plasticisers for use in automotive trim, fogging tests could be regarded as just an elaborate form of volatility test. Up to a point any accurate measurement of volatile loss from a plasticiser or a compound will give a good indication of its fogging propensity. However condensate collection takes much better account of the disproportionate effect of more volatile components. Any 'nonfogging' plasticiser must not only be intrinsically low in vapour pressure but needs to be consistently free of lower molecular weight impurities. This is discussed further in the context of product quality in Chapter 8.

Table 3.3 compares typical fogging values for a selection of widely used plasticisers having a broad range of vapour pressure. The values are only indicative of average values to be expected from materials of good commercial quality.

TABLE 3.3
TYPICAL FOGGING VALUES FOR VARIOUS PLASTICISERS

FOGGING VALUE	Molecular Weight	Vapour Pressure at 160°C (mbar)	Method A (%)	Method B (mg)
DOP	390	0.161	35	4.8
DIDP	447	0.039	76	0.6
Linear C9–11 phthalate	450 av.	0.022	90	0.35
Tri-2-ethylhexyl trimellitate	547	0.003	98	0.2

Despite the emphasis given to the fogging characteristics of plasticisers (not least by suppliers for reasons of marketing differentiation) the differences between plasticisers have often been swamped in practice by the effects of other compound ingredients, particularly liquid stabilisers. Analysis of 'real' fog on real windscreens has shown an enormous number of components among which plasticisers are by no means dominant. One such material of known origin is triethylene diamine[49] which is still commonly used as a catalyst in the production of flexible polyurethane foams for car seating despite having a molecular weight of only 112.

There has recently been a tendency to attach additional significance to fogging test results beyond their original purpose. A view is taken that the presence of any organic chemical vapour in the passenger compartment presents a potential health risk to the occupants. If the fog test is regarded as an indicator of the presence of such vapour then achievement of zero fogging becomes a target. However there has been no rational risk assessment to support this approach (see Chapter 9).

3.5   MIGRATION OF PLASTICISERS INTO LIQUIDS

3.5.1   EXTRACTION BY ORGANIC LIQUIDS

Any material having the correct polarity characteristics to be a PVC plasticiser is likely to be completely miscible with most organic liquids. Hence unlike the process of volatile loss where vapour factor is a limiting factor, the only constraint on the loss of plasticisers from PVC immersed in such liquids is normally their rate of migration to the surface. This is supported by the data on extraction of plasticisers by various liquids presented in Appendix 1. In contrast to volatile loss the figures show a very strong dependence on plasticiser level, this being characteristic of diffusion control. The results abstracted for Table 3.4 are typical in this respect. Details of the test method used are given in Appendix 1.

TABLE 3.4
EXTRACTION OF DOP FROM PVC – EFFECT OF PLASTICISER LEVEL

PLASTICISER LEVEL (p.h.r.)	Extraction by		
	Petrol g/m^2	Mineral oil g/m^2	Olive oil g/m^2
30	20	1.3	0.6
50	90	10	21
100	187	43	79

The greater extracting effect of 'petrol' (actually a 4:1 mixture of iso-octane with toluene) relative to the oils is due to much greater penetration of the PVC by its smaller more mobile molecules. Weighing the PVC samples immediately after immersion shows a considerable uptake of petrol which is allowed to evaporate before the final weighing and calculation of plasticiser weight loss. Some standardised tests (e.g. DIN 53 476) include the initial post immersion figure in the result. Absorption of the extractant results in a more open PVC structure facilitating diffusion as well as affecting the compatibility of the plasticiser with the polymer as discussed in 3.2. The swelling effect of more polar solvents is greater than for hydrocarbons and leads to even faster plasticiser loss.

Diethyl ether for example is capable of complete extraction of most plasticisers from PVC and it is used for this purpose in the quantitative analysis of plasticised compositions.

As with volatile loss, specifications for resistance to extracting liquids may be expressed in terms of maximum weight loss or change of physical properties. Articles with a thick cross section will obviously experience less change in plasticiser level and consequently less change in physical properties than thinner articles subjected to the same conditions. This needs to be taken into account in interpreting standard extraction data.

With the exception of triaryl phosphates the monomeric plasticisers are all susceptible to extraction by oils and solvents. For meeting specifications requiring extraction resistance it is conventional to use polyester plasticisers. Their good performance results from slow diffusion in line with high molecular weight. Because extraction is so dependent on plasticiser level the highest resistance is obtained for any plasticiser by using it at the lowest level consistent with the required end properties. Polyesters are frequently used in conjunction with other plasticisers which compensate for technical limitations such as poor low temperature performance. By careful formulation the benefits of the polyester's permanence are maintained even though the other plasticiser component may be intrinsically migratory.

The emphasis of the foregoing discussion has been on the effect of organic liquids on PVC products. Situations arise where the converse effect is of greater significance and adulteration of the extracting liquid by plasticiser is unacceptable. The most important case is food packaging where European legislation requires results using olive oil as a simulant for fatty and oily foods.

3.5.2 EXTRACTION BY WATER PLUS SURFACTANT

The effect of water in reducing the compatibility of plasticisers with PVC has been discussed. In the absence of surfactants the relative ease of removal of different plasticisers by water does not fit the most obvious expectation, that is increased resistance with increasingly hydrophobic molecular structure. The figures for soapy water extraction shown in Appendix 1 do, however conform to this expected pattern. If plasticised PVC is immersed in water containing surfactants these provide a mechanism for dispersion of the plasticiser as it arrives at the surface and progressive extraction can then occur. With very high concentrations of surfactant removal from the surface can become faster than diffusion through the PVC and the process becomes diffusion controlled.

The test used to obtain the soapy water extraction results in Appendix 1 differed from the method for measuring oil and petrol extraction, employing a much thinner sheet immersed in boiling soap solution. Results are quoted as % loss of compound weight rather than loss per unit area. This practice was consistent with the requirements of an important flexible PVC application at the time the test was devised, namely PVC baby pants which have since been eclipsed by the disposable nappy. The order of merit of the different plasticisers shows the benefits of high molecular weight and low polarity, the greatest resistance being shown by di-isotridecyl phthalate and the trimellitates. Polyester plasticisers as a class are not particularly resistant to soapy water, the most susceptible being the 'uncapped' types with hydrophillic end groups.

3.6   MIGRATION OF LIQUIDS INTO PLASTICISED PVC

3.6.1   SWELLING EFFECTS

As discussed in 3.4.1, immersion of a flexible PVC specimen in a solvent results in simultaneous absorption of solvent and loss of plasticiser from the surface into the solvent. The rate and extent of swelling of the specimen will depend respectively on the solvent's molecular size and polarity. On removal from the solvent the sample will be found to have undergone a net gain in weight and to have increased in softness as a result of plasticisation by the solvent. The length of time it remains in this state will depend on the volatility of the solvent. Contact with an involatile solvent or plasticiser will cause slower but more permanent swelling. The susceptibility of the PVC to swelling is dependent on its initial softness. It is well known that plasticised PVC is an unsuitable material for flooring or cables in areas where plasticiser spillages are likely to occur.

3.6.2   STAINING EFFECTS

The enhancement of plasticisation by absorption of a volatile solvent renders flexible PVC temporarily more susceptible to mechanical damage, soiling and staining. The solvent will act as a carrier for any soluble coloured material present. However staining of plasticised PVC commonly occurs without the intervention of an external solvent. Contact with any coloured material (or a colour precursor in which colour can be developed for example by exposure to light) which is soluble in plasticiser will result in staining. Since flexible PVC is to some extent water permeable this also makes it susceptible to staining by water soluble materials.

The main factor governing staining properties is the degree of plasticisation. Staining increases with increasing plasticiser level and tends to be worse for more efficient plasticisers. This is an obvious consequence of easier diffusion of the staining medium through a more open PVC structure. However the influence of plasticiser is more complex than this and it appears that the strength of interaction between the plasticiser and PVC is an important factor. Studies by Pinner and Massey[50] showed that low staining was obtained when the solubility parameter of the plasticiser closely matched that of PVC. This provides a rationale for the long established status as butyl benzyl phthalate as the standard stain resistant plasticiser for PVC flooring.

3.7 MIGRATION OF PLASTICISERS INTO SOLIDS

3.7.1 MIGRATION INTO SOLIDS – GENERAL

Plasticisers can be removed from PVC by contact with any absorbent material and the effects of rubbing with powdered minerals have been included in studies of plasticiser migration.[51] However migration into solids assumes most importance where the receiving material is another polymer. It is usually the effect on this polymer which is of concern rather than any change in properties of the PVC In situations where this can occur in practice the affected polymer can be a surface coating, a plastic article or component or an adhesive. The main effects of migration in these cases are:

   for coatings – softening and visible marring
   for structural components – loss of impact strength
   for adhesive bonds – softening and loss of adhesive strength.

Plasticisers show greater tendency to cause these defects the lower their molecular weight, the greater their linearity and the closer their polarity or solubility parameter matches that of the receiving polymer. The resistance of a polymer to plasticiser acceptance is increased by crystallinity and crosslinking. Table 3.5 gives a clear indication of the effect of plasticiser structure on migration. It shows the loss of various plasticisers from compounds initially containing 50 p.h.r of plasticiser into a flexible polyurethane foam.[88] The test conditions were as described in 3.7.2.

The results show the highest levels of migration associated with low molecular weight (DBP) and high linearity (DOA).

3.7.2 MIGRATION INTO SURFACE COATINGS

This topic provides a historical link with the early days of plasticiser technology in that the coating polymer with the greatest susceptibility to

TABLE 3.5
MIGRATION OF PLASTICISERS INTO POLYURETHANE FOAM

PLASTICISER	Molecular Weight	Migration (g/cm²)
DBP	278	0.40
DOP	390	0.17
DIDP	447	0.09
TOTM	547	0.03
DOA	371	0.33
BBP	312	0.22

migration was also the first material to be plasticised on any scale, that is nitrocellulose. Nowadays it has been superseded by other materials for most applications although nitrocellulose lacquers continue to be used, particularly for wood finishing. If a plasticised PVC article containing DOP remains in contact with a nitrocellulose lacquered surface for any length of time it is liable to adhere to it. Forced removal then damages the lacquer surface which shows signs of softening and swelling. Table 3.5 compares the marring of film cast from a nitrocellulose wood lacquer when held in contact with PVC containing phthalates and plasticisers of higher molecular weight.[52] The test procedure was based on ISO 177 with additional performance parameters included. Conditions were: temperature 70°C, contact pressure 13.8 mPa, duration 24 hours. Plasticiser levels were adjusted to give a low Shore hardness in order to accentuate differences in migration.

TABLE 3.6
MIGRATION INTO NITROCELLULOSE

PLASTICISER	p.h.r Initial	Shore A	Weight gain (g/m²)	Marring (qualitative)
DOP	88	56	30.3	Film greatly softened
DIDP	96	55	8.4	Sticky surface
Linear C9–11 phthalate	96	55	12.2	Sticky surface
Tri-2-ethylhexyl trimellitate	95	57	0	Nil
Linear C7–9 trimellitate	92	58	0	Very slight
Linear C8/10 trimellitate	106	57	0.2	Very slight
Polyester 1	95	57	Not measured	Film disintegrated
Polyester 2	95	54	Not measured	Film disintegrated

As might be expected from molecular weight and polarity considerations DOP showed a high level of migration and the trimellitates showed no effect. What may appear somewhat surprising is the poor performance of the polyesters here. It is presumed that the extreme marring observed resulted from migration of lower molecular weight fractions with relatively high mobility and a high degree of interaction with nitrocellulose.

The dominant position in car finishes once occupied by nitrocellulose paints is now occupied by thermoset acrylics. These are used for both the factory applied top coats and refinish topcoats and are based on acrylic polymers which are crosslinked in the cure process to give finishes which are resistant to damage by solvating materials. Table 3.7 shows the effect of the same plasticised compounds covered by the previous table when they were placed in contact with samples cut from a car body finished with a standard acrylic paint.

TABLE 3.7
MIGRATION INTO THERMOSET ACRYLIC CAR PAINT

PLASTICISER	Weight gain (g/m^2)	Swelling (10−2 mm)
DOP	13.6	1.5
DIDP	7.0	0.8
Linear C9−11 phthalate	10.1	0.8
Tri-2-ethylhexyl trimellitate	0.9	<0.5
Linear C7−9 trimellitate	1.6	<0.5
Linear C8/10 trimellitate	0.9	<0.5
Polyester 1	10.6	1.0
Polyester 2	0.9	<0.5

Here the swelling was far lower than with nitrocellulose although differences between the plasticisers were still apparent and marring was observable as a raised edge marking the contact area in the worst cases.

3.7.3 MIGRATION INTO THERMOPLASTICS

Polystyrene is particularly susceptible to marring by migrating plasticiser. The figures in Table 3.8 were again produced by the modified ISO method for the same set of compounds used to test the marring of nitrocellulose and acrylic coatings.

The relative performance of the different plasticisers shows clear structural influences. The greatest migration was given by DOP (lowest molecular weight of the group) and the linear C9−11 phthalate (most linear structure). DIDP with the same molecular weight and polarity as the latter showed higher migration resistance as a result of its more branched structure. The polyesters, even the one with lower molecular weight, proved to be migration resistant with polystyrene in contrast to their poor performance with nitrocellulose. This is consistent with a low

TABLE 3.8
MIGRATION INTO POLYSTYRENE

PLASTICISER	Weight gain (g/m^2)	Swelling $(10-2\,mm)$	Marring (qualitative)
DOP	25.3	3.0	severe
DIDP	11.6	1.5	moderate
Linear C9–11 phthalate	28.5	3.0	severe
Tri-2-ethylhexyl trimellitate	0.7	<0.5	very slight
Linear C7–9 trimellitate	1.9	<0.5	slight
Linear C8/10 trimellitate	1.1	<0.5	very slight
Polyester 1	0.7	<0.5	very slight
Polyester 2	−1.0	<0.5	nil

level of compatibility between the end groups and aliphatic ester repeat units of the polyesters and the aromatic hydrocarbon structure of polystyrene. Taken together with the results in the previous two tables the low migration shown by the trimellitates here is indicative of their broad spectrum of resistance to migration into polymers.

Most plastics including styrene copolymers show far lower absorption of migrating plasticisers than styrene homopolymers. For example when the foregoing test was repeated for the same compounds in contact with ABS, none of the ABS specimens gained more than $0.9\,gm^{-2}$, and all were free of measurable swelling and visible marring. However this is not to say that plasticiser migration cannot have serious consequences in such polymers. A very low level of absorption of plasticiser in a rigid thermoplastic subject to localised stress can result in the phenomenon of environmental stress cracking. Contact with plasticised PVC has been found to cause this problem in (for example) ABS, polycarbonates and rigid PVC.

As demonstrated by Hecker and Perry[14] and for reasons discussed in Chapter 1 unplasticised PVC is normally resistant to absorption of plasticiser at ambient temperatures. However absorption in a stressed area can be enough to have a large localised effect on tensile properties by antiplasticisation resulting in crack initiation and propagation.

Another example of low levels of plasticiser migration seriously impairing the performance of the receiving polymer can be found in the use of PVC as the sheathing material for the polyethylene primary insulation of high frequency cables. The intrinsically low dielectric loss factor of polyethylene is likely to be pushed above specification limits by contamination with plasticiser molecules containing responsive dipolar groups. The plasticiser structural factors providing resistance to this effect are high polarity (hence low interaction with PE), high molecular weight and low linearity. Selected types of polyester have been found to be satisfactory in this application.

Migration of plasticiser from one plasticised PVC composition to another is a common phenomenon which, because it seldom causes obvious problems is generally overlooked. The largest use of multilayer flexible PVC is in cushion vinyl flooring which is normally produced by successive spreading of four differently formulated plastisols starting with a non-woven glass fleece base. These are, in sequence an impregnating plastisol followed by an expandable layer containing a blowing agent, a clear top coat and a backing coat. Cadogan and McBriar[53] studied the transfer of plasticisers in three-layer simulated flooring samples using ^{14}C labelled plasticisers, obtaining the type of results shown in Figure 3.5 and Table 3.9.

The plastisols used to form the three layers each contained a different single plasticiser. After processing each layer contained all three plasticisers as shown. Other tests in the same study measured transfer of plasticisers at 70°C between layers produced separately and held in contact under pressure.

Overall the results showed that difference in plasticiser levels between the layers were not a necessary condition for migration to occur. The

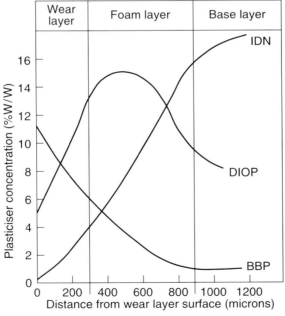

Fig. 3.5 Distribution of plasticisers within a sample of 3 layer cushion vinyl flooring.

TABLE 3.9

PLASTICISER CONCENTRATIONS IN THE THREE LAYERS OF A CUSHION VINYL
FLOORING SAMPLE BEFORE AND AFTER FUSION

	Plasticiser concentration (%W/W)					
	Wear layer		Foam layer		Base layer	
	Initial	Final	Initial	Final	Initial	Final
BBP	22.0	9.2		3.4		0.9
DIOP		9.9	28.0	12.5		9.0
IDN		1.5		8.0	25.4	16.8
TOTAL	22.0	20.6	28.0	23.9	25.4	26.7

fastest migration was shown by plasticisers of low molecular weight and/
or high linearity. The study leads to the conclusion that omission of a
particular plasticiser from any component of a multi-layer PVC product
for reason of a known end performance defect (e.g. staining, U.V.
resistance) may be rendered pointless if that plasticiser is included in one
of the other layers.

3.7.4 MIGRATION INTO ADHESIVES

Adhesive PVC sheet and tapes are used in a variety of applications where
migration of plasticiser into the adhesive layer can impair its function.
Plasticisation of the adhesive polymer results in a loss of adhesive strength
and this combined with the effect on the PVC of material loss can lead to
lateral shrinkage. The variables in an adhesive influencing its receptivity
to migration are the polarity of the polymer (in relation to the migrating
plasticiser), its molecular weight and the degree of crosslinking. Improvements in adhesives during the last 25 years have allowed the use in
adhesive PVC products of commodity phthalates which hitherto would
have been regarded as being too migratory.

For decorative sheet and labels the main adhesives used are high
molecular weight emulsion acrylics. These are used successfully on PVC
sheet containing various phthalates which from polarity considerations
would be expected to be quite compatible with the adhesive, including
DIDP, DOP and even the faster migrating linear phthalates. In these
applications the low plasticiser levels used (typically not much above
20 p.h.r.) no doubt act as a significant constraint on migration rate.

Acrylic adhesives are also used in medical tapes but here they are based
on relatively low molecular weight solution polymers in order to achieve
the required softness for skin adhesion. Medium molecular weight poly-

ester plasticisers are conventionally used in PVC for this application. In cases where total resistance to migration is required specialised grades of nitrile rubber are used as PVC plasticisers.

Electrical insulating tapes are coated with adhesives based on mixtures of natural and synthetic (hydrocarbon) rubbers, sometimes crosslinked. Except when intended for high temperature use the PVC is normally plasticised with a general purpose phthalate, DOP, DINP or DIDP. However, high temperature tapes for service up to 105°C require the use of a non-migratory plasticiser (most typically a polyester) to withstand the dual effects of accelerated migration into adhesive and volatile loss.

3.8   BARRIERS TO MIGRATION

Lacquer finishes have been used to improve the durability of PVC spread coated products for many years. Traditionally these have been solvent dispersed, the polymers being typically mixtures of vinyl chloride copoly-mers and poly(methyl methacrylate). Pressures to replace solvent based coatings have led to the development of alternative aqueous systems. The main objective of applying these coatings to PVC is to provide a surface with less tack and tendency to soiling than the soft PVC compositions which they cover. It is assumed that in order for this protection to be maintained during service the top coat must have long term resistance to plasticiser migration although comparative data on the performance of these coatings are not generally available.

In the PVC flooring industry the use of U.V. cured finishes is common-place in the USA but much less so in Europe. They are acrylate functional urethane systems which on curing give a crosslinked structure with high resistance to soiling and staining. Again it has to be assumed that relative impermeability to plasticiser migration is an inherent feature of such coatings. There is some debate in the industry concerning their physical permanence in practice.

It might be expected that these conventional anti-soiling coatings could be effective in reducing the volatile loss of plasticisers from PVC products but there is a lack of published information to demonstrate this. However patents exist which make specific claims for prevention of plasticiser migration by various types of coating including fluoropolymers,[54] cross-linkable functionalised PVC[55] and acrylic resins containing silane com-pounds.[56]

Whilst unplasticised PVC will absorb plasticiser at 70°C (a typical temperature used for migration testing), at lower temperature it demon-strates useful barrier properties against plasticiser migration. At one time it was common practice to formulate wear layer plastisols for cushion vinyl flooring in which a considerable proportion of the liquid content was volatile. This appeared to produced a superficial 'skin' not extending the full thickness of the wear layer and containing very little residual plasticiser. It was accepted that the persistence of this unplasticised skin during service was essential for the maintenance of stain resistance. An interesting variation on this theme is the observation that short immersion of a plasticised PVC sample in n-pentane (removing most of the plasticiser from the surface) followed by evaporation at 50°C for 15 minutes makes the bulk of the plasticiser in the article more resistant to extraction.[57] The use of stretched unplasticised PVC film as a barrier to plasticiser migration has also been claimed.[58]

As an alternative to the application of a separate barrier coat the recently developed technology of low pressure plasma treatment appears to have interesting possibilities for improving plasticiser retention. The plasma consists of an ionised gas which can participate in reactions with the surface of an object of any shape to bring about some desired modification. Oxygen plasma treatment is now a well established method of surface activation of components made from inert plastics to facilitate painting or adhesion. In these cases its main function is to provide surface reaction sites by oxidation. Plasma processes can also be used to induce surface crosslinking and it is this which is of potential value in PVC as a barrier to plasticiser migration. It has been reported that plasma treatment using a wide variety of gases are effective in reducing plasticiser volatility.[59] Specific applications of uses mentioned in patents are roof membranes with resistance to bitumen migration[60] and PVC tubing resistant to hexane extraction at its interior surface.[61] The second of these is interesting as an example of non-planar barriers obtainable by this treatment. However the use of this technology for improving the performance of plasticised PVC is still in its infancy.

Development of effective barrier technology for use with low cost PVC compositions could provide appropriate alternatives to more expensive migration-free materials for some applications. However it would involve additional process cost and complexity and could sacrifice some of the beneficial aspects of plasticised PVC surfaces, a general example being their weldability and a more specific example being their biocompatibility in medical devices.

3.9 THERMO-OXIDATIVE EFFECTS

The various processes of plasticiser loss considered so far in this chapter simply involve separation of plasticiser from the polymer, neither being involved in any chemical reaction. this and the following sections deal with processes in which deterioration in the properties of the plasticised composition originate from some chemical change in the plasticiser. Such processes may or may not involve the polymer in chemical reactions. In the case of plasticiser oxidation whether thermally or photo-induced, it is subsequent changes in the polymer which have the most serious consequences.

The type of plasticiser used can have a major impact on the stability of the polymer. Before expanding on this it is worthwhile distinguishing between the various criteria by which the thermal stability of plasticised PVC may be assessed. These are:

(a) colour development
(b) hydrogen chloride release
(c) deterioration in physical properties.

Use of different criteria will be appropriate to different applications. Whilst the mechanisms of all three processes are inter-related the observed effects often do not run parallel. Thermal stability is determined by accelerated tests carried out at temperatures far in excess of those encountered by the end product in service. Test temperatures are often similar to those used for processing the compound although the duration is extended well beyond normal process times.

Colour development is a result of the formation of long sequences of conjugated double bonds (polyenes) as hydrogen chloride is lost, the longer the sequence the darker the colour. However colour is indicative of the level of dehydrochlorination only so long as no further reactions occur to shorten the polyene sequences and thereby destroy their chromophoric effect. The most probable such reaction is crosslinking between polyene sequences in adjacent polymer chains.

Hydrogen chloride release is detected when the capacity of the stabiliser system for HCl binding becomes exhausted. Tests for HCl release of PVC insulating compounds are used by the cable industry and are of direct significance here because of the potential for corrosion of electrical components. The tests measure the induction time before HCl is detected by indicator paper held above the specimen.

Change of mechanical properties has already been discussed in the context of volatile loss. It is evident that plasticiser loss increases hardness and stiffness. However it does not follow that the final physical properties then match those of an unaged reference compound with the same plasticiser content as the heat aged composition since it will have reached this point by an entirely different heat history and its internal structure is therefore likely to be different. The additional changes in physical properties superimposed on those due to volatile loss result from reactions accompanying dehydrochlorination.

Table 3.10 shows the results of a 7 day 100°C heat ageing test in an oven with fast linear air flow (2 m/s). The compounds tested contained 50 p.h.r. of plasticiser and were stabilised with 5 p.h.r. of dibasic lead phthalate. Where antioxidant is indicated this was present at 0.5% on plasticiser. The plasticisers were a selection of commercial dialkyl phthalates.

TABLE 3.10
HEAT AGEING OF PHTHALATE COMPOUNDS +/− ANTIOXIDANT

PLASTICISER	Volatile loss (% plasticiser)	Retention of elongation (%)	Retention of tensile strength (%)
DOP	46.6	61	93
DOP + antioxidant	43.8	62	94
DIOP	39.8	2	7
DIOP + antioxidant	39.5	64	93
L79P	27.1	88	89
DIDP	13.7	38	65
DIDP + antioxidant	10.3	94	95

Determination of retention of tensile properties is subject to considerable error and too much significance should not be attached to small differences. In broad terms the main features of the results are clear:

(a) In the absence of antioxidant there was a wide variation of retention of elongation and tensile strength unrelated to volatility differences.
(b) The greatest loss of tensile properties was given by the phthalates with the most branched alkyl structures.
(c) Inclusion of antioxidant had little effect on volatility but corrected poor retention of tensile properties. The order of retained elongation then fell into line with volatile loss, DIDP showing virtually no change.

The relationship between carbon chain branching and susceptibility to oxidation is well understood from classical studies on the oxidation of hydrocarbons. Branching points provide favoured oxidation sites because of the relative ease of removal of the attached tertiary hydrogens. Addition of an oxygen molecule at this point forms peroxy free radicals. These can then abstract hydrogen atoms from PVC chains thereby initiating dehydrochlorination as shown in Figure 3.6.

The results in Table 3.10 differentiate between plasticisers with different oxidative stability and show that under the conditions of the test the less

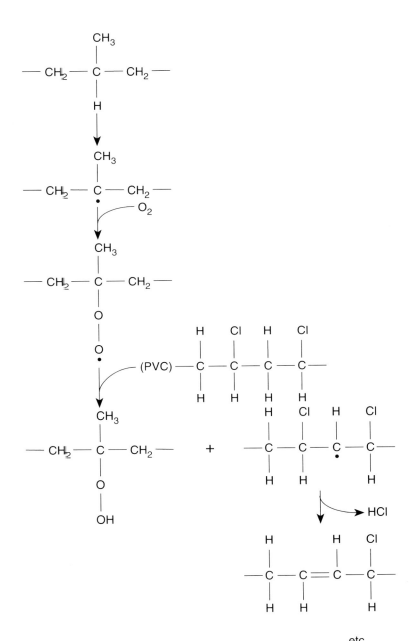

Fig. 3.6 Initiation of PVC degradation by free radical transfer from plasticiser.

branched phthalates (DOP and L79P) do not require antioxidant. However as test temperatures are progressively increased the need for antioxidant becomes more general. Some plasticisers demonstrate intrinsic oxidation resistance even at the highest test temperatures normally used for PVC (for example tri-heptyl trimellitate, see Table 3.11). Even so the inclusion of an antioxidant in compositions used at these temperatures is always advisable since it guards against the initiation of degradation via some other oxidative weak link in the formulation. Non-volatile plasticisers marketed for high temperature PVC applications are nearly always supplied containing dissolved antioxidants (see Chapter 2, 2.4.8).

Table 3.11 shows the results of 7 day heat ageing tests in the oven previously described but this time at 136°C. The test compounds were plasticised with various linear and branched trimellitates at levels adjusted to give a standard softness. The plasticisers were tested with and without 0.5% of a high molecular weight hindered phenol antioxidant (Topanol CA–I.C.I.). Where tests were repeated on different occasions the spread of results obtained is indicated. The only plasticiser apparently free from oxidative degradation effects without antioxidant was the trimellitate of n-heptanol. Those with branched or longer linear chains required the presence of antioxidant to achieve the same performance.

TABLE 3.11
HEAT AGEING OF TRIMELLITATES +/− ANTIOXIDANT

TRIMELLITATE	p.h.r.	Antioxidant	Retention of elongation (%)
n-heptyl	45	−	88–103
n-heptyl	45	+	103
C7–C9 (80% linear)	47	−	52–56
C7–C9 (80% linear)	47	+	99
C7–C9 (50% linear)	47	−	51–61
C7–C9 (50% linear)	47	+	98
2-ethylhexyl	50	−	58
2-ethylhexyl	50	+	88
C8–C10 (100% linear)	53	−	53–60
C8–C10 (100% linear)	53	+	98
C9–C11 (80% linear)	64	−	44–49
C9–C11 (80% linear)	64	+	101

Oxygen attack on alkyl groups in the plasticiser molecule provides the dominant destabilising interaction between plasticisers and PVC for phthalates, trimellitates and other alkyl esters in tests relating to service at elevated temperatures. At temperatures above about 200°C fragmentation of the plasticiser molecule begins to occur. Boiling DOP (boiling point 386°C) for 1 hour has been found to result in 30 % decomposition mainly into phthalic anhydride, 3-methylene heptene and 2-ethylhexanol in the approximate molar ratio 1:1:1[62] indicating the following overall decomposition reaction:

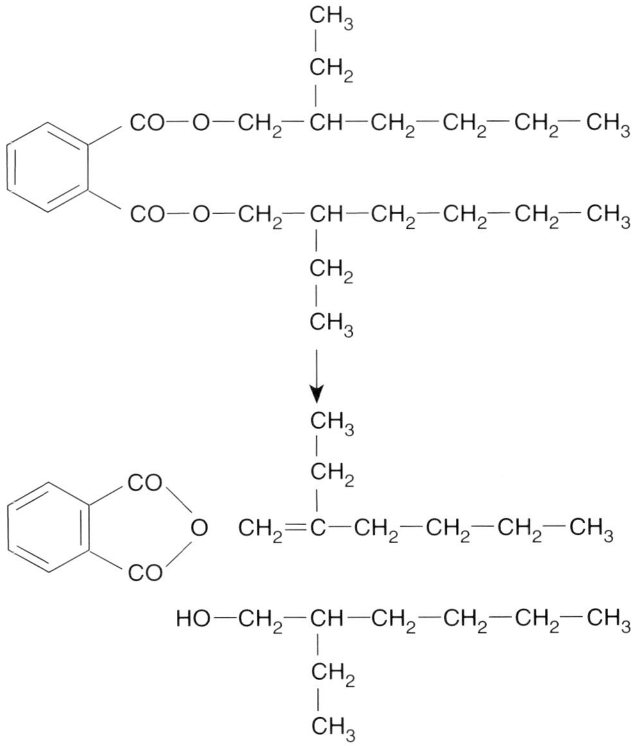

It was found that acid catalysis lowered the temperature of decomposition and altered its course, resulting in the formation of phthalic acid rather than phthalic anhydride plus an additional mole of 3-methylene heptene in the place of 2-ethylhexanol. These products were found after heating DOP with 5% of PVC for 40 hours at 170°C, the PVC here acting as the acid source. The same treatment with DOS produced sebacic acid and 3-methylene heptene.

It is perhaps unlikely that these reactions would influence the stability of PVC compositions under service conditions. However the nature of the products of thermal decomposition of plasticisers can be of significance in assessment of HSE risks in plastisol processes. Recirculation of process fumes in gas heated ovens can result in the presence of plasticiser decomposition products in the fumes and their condensates.

With other types of plasticiser different mechanisms of decomposition may impact on PVC stability. Chlorinated n-paraffins have a structural resemblance to PVC although lacking the regularity of the polymer's repeat units. Like PVC they decompose on heating to give hydrogen chloride and they are responsive to stabilisation by various materials used to stabilise PVC. Relative to most ester plasticisers their presence has a slight destabilising effect on PVC.

3.10 PHOTOSTABILITY AND WEATHERING

Since the damaging effects of light are most likely to occur in outdoor applications, photostability of PVC is most often covered in the context of weathering resistance. However real weathering involves a variety of interacting factors in addition to photodegradation –

- thermo-oxidative reactions
- volatile loss
- aqueous extraction and displacement
- chemical reactions with water
- chemical reactions with airborne contaminants

Combination of these factors with formulation variables – additive and polymer types, ratios and interactions provides an almost limitless field for study. Although guideline data exist in raw material suppliers' publications and in the technical literature, development of new formulations for outdoor products always requires a prudent approach with prolonged testing under conditions known to simulate the service environment. The criteria by which weathering performance is judged include discolouration (which may be uniform or spotty), loss of gloss, microcrazing, exudation, 'chalking' and deterioration in mechanical properties.

Whilst the weathering performance of a complex formulation might not be fully predictable, dehydrochlorination studies on simple compositions containing carboxylic ester plasticisers with varying branching and aromatic content have produced a clear and consistent explanation of the relationship between structure and photostability.[63, 64] The technique used in this work was laser-Raman spectroscopy which is capable of measuring very low levels of dehydrochlorination with high spatial resolution and at the same time detecting the onset of PVC–PVC crosslinking. The interaction with PVC is similar to that in thermo-oxidative degradation in that free radical reactions start with the plasticiser and then transfer to the polymer. However the necessary first stage is the absorption of light energy and it is here that the first plasticiser structural effect comes into play. The postulated sequence of events is described below.

(a) The shortest wavelength of radiation reaching the earth's surface is 292 λm. It is the relative absorbtion by different plasticisers of U.V. at wavelengths greater than this which is of significance in relation to their weathering performance. Figure 3.7 compares the absorbance of a number of plasticisers. Increasing aromatic content in the sequence DOA = DOS < DOP < BBP shifts the absorption curve to higher wavelengths. An additional structural effect accounts for the high absorbance of TOTM relative to DOP this being a consequence of the additional carboxyl group attached to the benzene ring.

(b) The absorbed energy leads to bond fracture and free radical formation involving oxygen. As with thermal oxidation this step is influenced by alkyl chain branching and length. Table 3.12 shows the levels of peroxides formed in samples of (uncompounded) linear and branched phthalates irradiated with a mercury vapour lamp for two weeks.

TABLE 3.12
PEROXIDE LEVELS IN IRRADIATED PHTHALATES

PHTHALATE TYPE	Peroxide level (µg/g)	
	before irradiation	after irradiation
LINEAR		
C6	<5	10
C7–C9	<5	12
C8–C10	<5	10
C7–C11	8	16
C9–C11	<5	18
BRANCHED		
C8 (DOP)	10	26
C9	10	28
C10	18	120
C13	17	250

(c) Plasticiser free radicals react with the polymer initiating dehydro-chlorination 'unzipping' reactions which lead to colour formation.

(d) Crosslinks form between polyene sequences in adjacent polymer chains. This reduces colour thereby disguising dehydrochlorination but also alters physical properties and reduces the compatibility of the plasticiser (and other additives) with the PVC increasing the probability of exudation. Any resultant increase in plasticiser

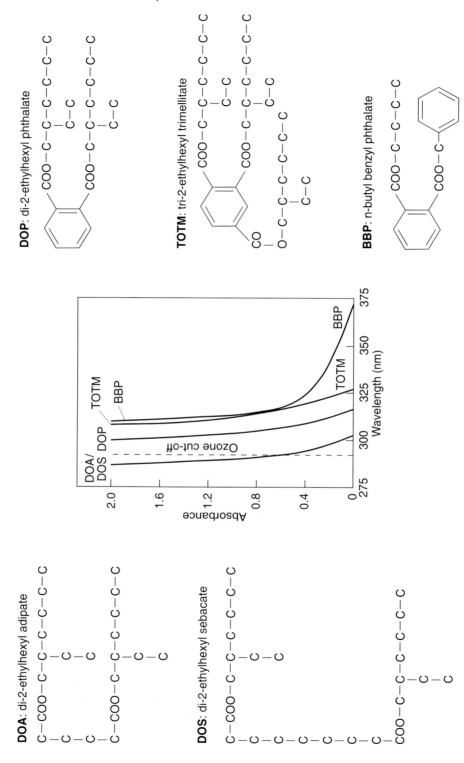

Fig. 3.7   Effect of plasticiser structure on UV absorption.

concentration at the surface increases UV absorption thereby accelerating the whole process.

Table 3.13 shows the levels of dehydrochlorination measured for samples of PVC exposed for 1 year on 45 degree weathering racks in three locations. The test formulations contained suspension PVC plus plasticiser (60 p.h.r.) dibasic lead phthalate heat stabiliser (3 p.h.r.) and stearyl alcohol lubricant (0.5 p.h.r).

TABLE 3.13

DEHYDROCHLORINATION OF WEATHERED PVC – EFFECT OF PLASTICISER TYPE

PLASTICISER	% Dehydrochlorination		
	Wales	Arizona	Queensland
ALIPHATIC ESTERS: Di-2-ethylhexyl adipate Di-2-ethylhexyl sebacate	0.009 0.004	0.005 0.004	0.003 0.003
LINEAR PHTHALATES: C6 C7–C9 mixed C7–C11 mixed C8–C10 mixed C9–C11 mixed		0.024 0.013 0.012 0.014 0.012	
BRANCHED PHTHALATES: C8 (DOP) C9 C10 C13		0.022 0.055 0.38* 0.51*	
OTHER PLASTICISERS Butyl benzyl phthalate Tri-2-ethylhexyl trimellitate	0.48* 0.033	>5* 0.21	>5* >5*

(*Raman spectra indicated crosslinking)

The influence of plasticiser level on the dehydrochlorination of DOP-plasticised compounds exposed in Queensland and Arizona for 1 year is shown in Fig. 3.8. Interestingly the increase of HCl loss with increasing plasticiser level runs counter to the effect of DOP on the thermal stability of PVC compounds heated under nitrogen. There not only does the plasticiser play no part in initiating HCl loss but on the contrary has a dramatic stabilising effect.[65]

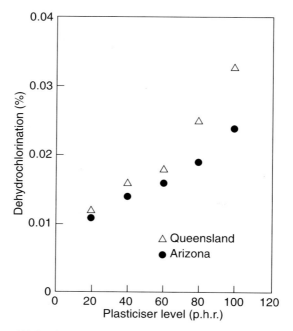

Fig. 3.8 Effect of DOP level on HC1 loss from weathered PVC.

3.11   CHEMICAL DEGRADATION OF PLASTICISERS

Under normal conditions of service most plasticisers are usually chemically stable. However most of them are esters and it is usually the ester group which provides the weak link in the molecule since it is susceptible to varying degrees to hydrolysis under alkaline or strongly acidic conditions.

e.g.

$$
\begin{array}{c}
CO-O-C_8H_{17} \\
| \\
(CH_2)_4 \\
|\\
CO-O-C_8H_{17}
\end{array}
\;+\;2\,H_2O
\;\xrightarrow{\;H^+ \text{ or } OH^-\;}\;
\begin{array}{l}
2\,C_8H_{17}OH \quad \text{octanol} \\
+ \\
CO-O-OH \\
| \\
(CH_2)_4 \quad\quad \text{adipic acid} \\
| \\
CO-O-OH
\end{array}
$$

dioctyl adipate

In practice it is alkaline hydrolysis which is far more likely to cause problems. In the presence of sufficient base the acid produced is converted to the salt and the equilibrium is displaced so that hydrolysis proceeds to completion (saponification) e.g.

dioctyl phthalate　　　　　calcium phthalate　　　octanol

This last reaction can occur in a number of different circumstances:

(a) Plasticiser spillage on concrete
Freshly set concrete contains a high proportion of calcium hydroxide and is highly alkaline (up to pH 11 or 12). In the course of time this is reduced by reaction with atmospheric carbon dioxide. It has been found that spillage of ester plasticisers on concrete can lead to disintegration of the surface. Because of their low volatility plasticisers are liable to remain in contact with the concrete for a long time and there is the potential for calcium phthalate formation to destroy its integrity to a considerable depth. For this reason it is usually recommended that unprotected concrete is not used for bunds or supports with plasticiser storage tanks.

(b) Contact between plasticised PVC and concrete
This occurs when flexible PVC floorcovering or PVC-backed carpet tiles are laid directly onto concrete flooring. This sometimes happens in new buildings when the surface is likely to be still at the highly alkaline stage. The usual sign of interaction is the obvious smell of an alcohol (e.g. 2-ethylhexanol if the plasticiser is DOP) which can reach an objectionable level. Saponification of plasticiser at the PVC surface will result in some depletion although this may be insignificant.

(c) Saponification by formulation ingredients
Occasionally it happens that a plasticised PVC product is found to have a strong alcoholic odour after manufacture and analysis then confirms the presence of the alcohol from which the plasticiser is derived. This is even though the plasticiser itself was probably virtually alcohol free and odourless prior to compounding. In this situation promotion of hydrolysis during processing by other ingredients of the formulation can be suspected. An example occurs in the use of calcium oxide as a desiccant in plastisols to remove traces of water which could otherwise cause bubbles or voids in the final product. The product of the desiccation reaction is calcium hydroxide. In the presence of additional water this has been known to catalyse hydrolysis under some conditions. Hence this desiccant needs to be used with caution.

The relative susceptibility of different esters to hydrolysis is influenced by steric factors. In general linear structures provide less steric shielding of the ester group from hydrolytic attack than more branched structures. Hence the order of hydrolytic stability of some common types of plasticiser is:

triaryl phosphates > branched phthalates > linear phthalates > adipates.

3.12   BIOLOGICAL DEGRADATION OF PLASTICISERS

Of the major thermoplastic materials, flexible PVC is the one most susceptible to microbiological attack. This is not due to the polymer which is biologically stable but to the nature and level of the additives used. Of these it is generally the plasticiser which is the most significant variable. As with chemical degradation the ester group normally provides the weak link for initiation of breakdown. The most significant degradative organisms are fungi of many different types. Fungi have been shown to produce esterases, the enzymes capable of breaking ester linkages. The fragment molecules so produced can then be metabolised by both fungi and bacteria.[66]

Biological degradation of plasticisers occurs on or close to the PVC surface and the degradation products are lost to the surrounding environment. The process continues with growth of the organism and diffusion of plasticiser to replenish the surface. Losses can become high enough to cause significant loss of flexibility but before this happens objectionable side effects may become apparent. Staining, commonly pink in colour is a result of bacterial activity. The familiar odours of mould and mildew growth may also occur as well as that of plasticiser decomposition products.

Any application in which plasticised PVC experiences prolonged exposure to wet or humid conditions, particularly outdoors or involving soil contact has potential for microbiological attack. For such applications the use of biocidal additives is standard practice (see Chapter 2, 2.4.10). In the absence of these biostabilisers there are significant differences in stability between different plasticisers. Non-esters have the inherent advantage of freedom from the influence of fungal esterases. The rough order of merit is similar to that for resistance to chemical hydrolysis:

HIGH RESISTANCE   chlorparaffins
polyethers (used in nitrile rubber)
phosphates (triaryl better than alkylaryl)
phthalates (branched better than linear)

adipates
sebacates
polyesters
LOW RESISTANCE   epoxy oils and epoxy esters

3.13   EFFECT OF PLASTICISERS ON FLAMMABILITY AND SMOKE GENERATION
(See also Chapter 2, 2.4.6 – Flame retardants and smoke suppressants)

In assessing the suitability of a material for an application involving fire risk, various aspects of its performance have to be taken into account – ease of ignition, spread of flame, heat release and the generation of smoke and toxic gases. Excessive emphasis on a single aspect can give an unrealistic view of overall risk relative to other materials. With the growth of the science and technology of fire testing this point has been widely accepted leading to greater questioning of the use of selective information to promote sectional interests.

Unplasticised PVC has the greatest resistance to burning of any commodity thermoplastic and is only combustible in extreme fire conditions when high temperatures have been reached. This fire resistance is a result of its high chlorine content (56.8%) and its mode of thermal decomposition which releases hydrogen chloride rather than flammable pyrolysis products. Not only is HCl not flammable but it has a positive flame retardant effect by participating in termination of free radical combustion reactions. However the fire performance of PVC is altered by the presence of organic additives, in particular plasticisers because of their high levels of incorporation. All plasticisers detract from the fire resistance of PVC to varying degrees, even those classified as 'fire retardant'.

Figure 3.9 shows the effect of plasticiser type and level on the oxygen index of PVC compounds made to the following formulation:

Suspension PVC, ISO viscosity number 125   100
Basic lead carbonate   3
Calcium stearate   0.5

Oxygen index is a test which is widely used for screening and quality control purposes. Specimens of the material are ignited with a flame in an oxygen enriched atmosphere. The concentration of oxygen below which they are found to be self extinguishing after removal of the flame is termed the oxygen index of the material. The test has the merits of simplicity and low cost but cannot be used to predict the performance of a material against specified application standards or in real fire situations.

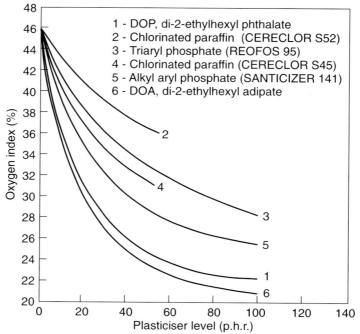

Fig. 3.9 Dependence of oxygen index of PVC on plasticiser content.

In Fig. 3.9 the greatest reduction in oxygen index is shown by DOP and DOA, neither of which has any capability for participating in flame retardant mechanisms. The slightly worse effect of DOA is presumably a result of its wholly aliphatic structure with higher hydrogen content providing greater fuel value. The same factor provides an explanation for the lower oxygen index of alkylaryl phosphates relative to triaryl phosphates.

The oxygen index of chlorparaffin plasticisers increases with increasing chlorine content. Whilst chlorparaffins provide low flammability at low cost their limited compatibility normally prevents their use as sole plasticiser. Phosphates (particularly triaryl phosphates) have high extender tolerance and therefore it is possible to devise a compatible flame retardant formulation with high plasticiser content using a combination of phosphate plus chlorparaffin. Such mixtures have been used successfully for many years to meet severe specifications limiting spread of flame in coal mine conveyor belting. Because of the poor low temperature flex of triaryl phosphates it may be necessary to include in the formulation a further plasticiser which will inevitably lower the flame resistance. As a general rule it is difficult to reconcile flame retardance and cold flex simply by plasticiser selection. Unless a transparent product is required it is unlikely that plasticiser selection alone would be relied upon to meet flame retardant specifications. Inorganic additives such as antimony trioxide or alumina trihydrate can be used as alternatives or in conjunction with flame retardant plasticisers.

The limitations of oxygen index as a predictor of performance in fire tests are illustrated by the results shown in Table 3.13. Two formulations having the same oxygen index were subjected to three standard spread of flame tests.[67] Formulation A contained a high proportion of triaryl phosphate with no additional flame retardant. Formulation B which contained antimony oxide and a reduced proportion of phosphate achieved the same oxygen index but showed much worse performance in the spread of flame tests.

TABLE 3.13
COMPARISON OF PERFORMANCE BY DIFFERENT FLAMMABILITY TEST METHODS

PVC	A	B
	100	100
Linear phthalate	17	67
Isopropylated phenyl phosphate	50	–
Antimony trioxide	–	12
Epoxidised soya bean oil	5	5
Ba/Zn stabiliser	3.3	3.3
Oxygen index (3.2 mm test sample) %	29.5	29.5
BS 3424 method 17 (0.18 mm unsupported film) burn length (mm) burn time (sec)	134 16.8	171 21.1
BS 3424 method 17 (900 g.m-2 coated fabric) burn length (mm) burn time (sec)	38 1.2	88 38.2
NF-P92-503 (900 g.m-2 coated fabric) Bruleur electrique classification	M2	M4

In addition to increasing flammability the incorporation of plasticisers tends to increase the level of smoke generated in fires. Smoke causes the hazard of obscuration of vision. It consists of solid particulate matter plus condensed droplets of liquid suspended in a mixture of air and gaseous combustion products. Under conditions giving incomplete combustion evaporation of plasticiser will contribute droplets of condensate and solid

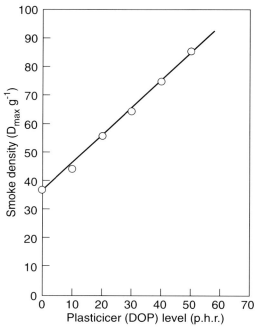

Fig. 3.10 Effect of DOP content on smoke generation from burning PVC. (With kind permission from Butterworth-Heinemann Journals, Elsevier Science Ltd.)

particulates from partial combustion in the gas phase. Greater aromatic content in the plasticiser structure favours the formation of the carbon skeletons which make up particles.

Figure 3.10 shows the relationship between smoke generation and DOP content.[68]

Despite the negative effects of plasticisers on the fire properties of PVC flexible PVC still shows better fire resistance at lower cost than most competitive materials. Consequently it remains the preferred option in most of its established applications where this factor is relevant. The negative feature of PVC, plasticised or unplasticised, in fires is its obvious propensity for hydrogen chloride generation. A further concern has been its potential for generation of dioxins under conditions of incomplete combustion. Further reference to these points is made in Chapters 6 and 9.

CHAPTER 4

The Phthalate Plasticisers

4.1 INTRODUCTION

The phthalate plasticisers are esters of ortho-phthalic acid:

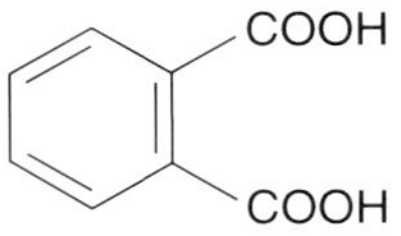

However they are not manufactured from the acid which is not an industrial commodity but from phthalic anhydride:

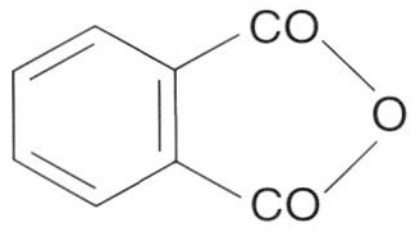

This chapter deals with the various commercial esters derived from phthalic anhydride and alcohols by straightforward esterification processes:

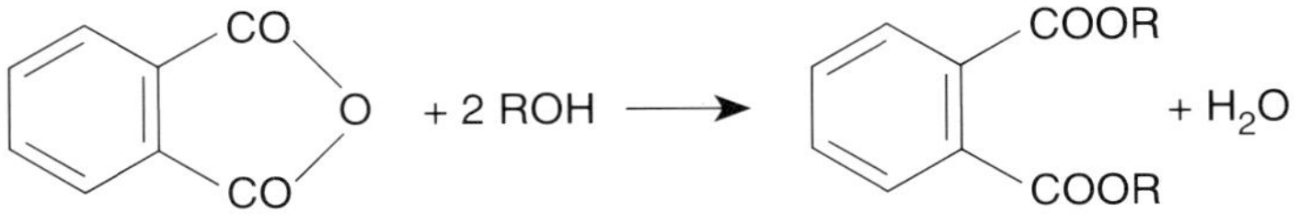

The benzyl alkyl phthalates which are manufactured by a different two stage process are covered at the end of the chapter.

In the above equation ROH may be a single chemical entity (for example 2-ethylhexanol), a mixture of isomeric alcohols with the same number of carbon atoms (e.g. isodecanol) or a mixture of alcohols with a range of carbon numbers (e.g. Linevol® 911, a mixed C_9, C_{10} and C_{11} alcohol consisting of linear and branched isomers). For purposes of comparison mixed phthalates of the latter type are classified according to their average carbon number. On this basis Linevol 911 phthalate can be regarded as a C_{10} phthalate.

Phthalates have a share of the total world plasticiser market of around 86% and many large plasticiser purchasers use only phthalates. This dominant position sometimes results in 'phthalate' and 'plasticiser' being used almost synonymously.

The phthalates used industrially range from C_1 dimethyl phthalate to C_{16}/C_{18} cetyl stearyl phthalate. However at least 90% of production and use is of phthalates in the narrow band between C_8 and C_{10}. Some chain lengths hardly appear in commercial use, not necessarily because they would not be technically useful but because of the absence of economical routes to the feedstock alcohols. The European tonnage distribution of phthalates of different carbon number is indicated in Figure 4.1 (some of

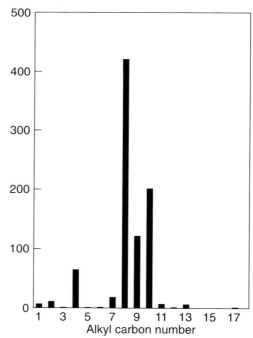

Fig. 4.1 West European consumption of phthalates (1994) according to size of alkyl chain.

the indicated tonnages are estimated values for illustrative purposes only). Mixed phthalates are included according to average chain length. Principle uses of the different phthalates are then shown in Table 4.1.

Phthalates of low molecular weight are too volatile to be used as PVC

TABLE 4.1
DIALKYL PHTHALATES – PRINCIPLE USES BY ALKYL CARBON
NUMBER

CARBON NUMBER	Main use
1	Solvent for polyester resin cure initiators
2	Plasticiser for cellulose acetate
3	Unknown commercially
4	Plasticisers for polyvinyl acetate
5	Small use in PVC
6	Relatively unknown commercially
7	Fast processing plasticiser for PVC
8 to 10	Main group of PVC plasticisers
11 to 13	PVC plasticisers with very low volatility
14 to 16	Unknown commercially
17	Processing lubricant for rigid PVC

plasticisers whilst those of high molecular weight are excluded by low compatibility. C_4 phthalates are used in significant quantities in PVC where they contribute fast fusion at low cost. However their volatility is far too high for most applications. No phthalate higher than C_{13} has sufficiently good processability or compatibility with PVC to be a useful plasticiser.

The pre-eminence of phthalates in the PVC industry is due to their cost effectiveness. Throughout most of their industrial history there has been a ready availability of low priced phthalic anhydride and alcohol feed-stocks. Simply from a technical viewpoint the all round performance of the C_8 to C_{10} phthalates has seldom been matched let alone bettered by any other type of plasticiser. Much of the research into PVC – plasticiser interaction has taken phthalates (particularly the chemically simple arche-type DOP) as the models and sought to explain what it is about their structure which makes them so effective. Taking a more empirical approach, the relationships between alkyl chain structure and technical performance of the phthalates are well defined as a result of a wealth of data on many examples.

As indicated in Chapter 1 the industrial status of phthalates and their range of uses has caused this group of materials to receive particular attention in relation to their toxicological and environmental hazards. As a result of this focus there has been some speculation about their long term future use in some applications. However, despite more than a decade of continuous attention no firm reason has been found to restrict sig-nificantly the use of phthalates and their worldwide use continues to grow.

4.2   PHTHALIC ANHYDRIDE

Phthalic anhydride (PA) is a basic commodity chemical with about 15 European producers having total capacity of around 800 kt/a. Most of this capacity is owned by producers of phthalate esters, these products constituting by far the largest use of PA.

TABLE 4.2
PHTHALIC ANHYDRIDE APPLICATIONS

Phthalate esters (plasticisers)	70%
Unsaturated polyester resins (for GRP)	15%
Alkyd resins (for paints)	13.5%
Miscellaneous uses (mainly colourants)	1.5%

PA is produced by catalytic air oxidation of either (a) naphthalene obtained by coal tar distillation or (b) orthoxylene.

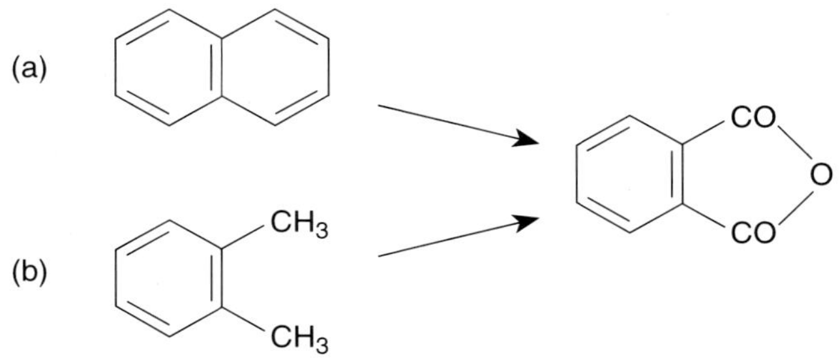

Whilst (a), the original process for manufacturing PA remains in limited use, the great majority is produced by route (b). Orthoxylene is a commodity obtained by distillation of the mixed C_8 aromatic products of refinery catalytic reforming plants. Factors influencing supply are capacity of the large specialised distillation units and competing demand for **para**xylene which can be produced by isomerisation of orthoxylene. Paraxylene is the feedstock for terephthalic acid, the principal raw material for polyester fibres and PET bottles.

4.3   PHTHALATE PRODUCTION PROCESSES

The main reactions occurring in phthalate production processes are:

Formation of monoalkyl phthalate occurs quickly at relatively low temperature by ring opening of phthalic anhydride. Conversion to the diester is slower and requires temperatures in excess of 140°C and the presence of a catalyst. The equilibrium reaction is driven to complete phthalate formation by removal of water, normally under vacuum. Excess alcohol used to ensure complete conversion of the PA is recycled. Smaller and more flexible phthalate plants use batch processes whereas the largest plants designed for dedicated manufacture of a single product can employ continuous process technology.

The catalysts used in phthalate production processes are either strong acids or materials which function by an amphoteric mechanism. Sulphuric acid and para-toluenesulphonic acid are the best known acid catalysts although methane sulphonic acid is a newer alternative claimed to give some process advantages.[69] Typical temperatures for acid catalysed processes are in the region of 150°C. The most commonly used amphoteric catalysts are titanates, for example tertiary butyl titanate. Reaction temperatures used in amphoteric processes are much higher than with acid catalysts, typically 220°C. Hence amphoteric catalysis is not suited to the esterification of phthalic anhydride with low molecular weight alcohols having low boiling points.

The esterification stage is followed by various purification treatments to remove unconverted raw materials, catalyst and any products of side reactions. The range of process variations can result in minor differences in trace impurities in the same phthalate from different suppliers although all may be equally capable of meeting the same sales/purchasing specifications. For most end uses such differences are seldom of technical significance.

4.4 C_1 AND C_2 PHTHALATES

Manufacturer references (see Appendix 1): 6, 8, 12, 14, 21, 36, 43

Dimethyl phthalate and diethyl phthalate, sometimes termed 'the lower phthalates' are produced from the commodity alcohols methanol and ethanol respectively. Because of the high volatility and water solubility of these alcohols manufacture of the phthalates requires the use of plants with specialised features.

Compared with (e.g.) DOP the number of producers of lower phthalates is very limited.

4.4.1 DIMETHYL PHTHALATE (DMP)

COOCH$_3$

COOCH$_3$

Viscosity at 20°C 18 mPa.s
(c.f. DOP 78)
Vapour pressure at 100°C 1.29 mbar
(c.f. DOP 0.001)

DMP is used as a plasticiser for cellulose acetate but only for small specialised applications. It is also used as plasticiser/ coalescing solvent in some polyvinyl acetate emulsions.

The main application of DMP is not as a plasticiser at all but as a phlegmatising solvent for MEK peroxide which is the main cold cure initiator for crosslinking of unsaturated polyester resins. Here it functions both as an inert reaction medium in the peroxide production process and dilutes the activity of the final product to allow safe transportation and handling. Typically this initiator is supplied in the form of a 50% solution in and is used at around 2% of the resin weight. Hence the final cured product contains of the order of 1% of DMP held in the structure by solvation and having no significant effect on properties.

An unusual and specific application of DMP is as an active ingredient of mosquito repellants.

4.4.2 DIETHYL PHTHALATE (DEP)

COOCH$_2$CH$_3$

COOCH$_2$CH$_3$

Viscosity at 20°C 13 mPa.s
(c.f. DOP 78)
Vapour pressure at 100°C 0.825 mbar
(c.f. DOP 0.001)

DEP is the main plasticiser used with cellulose acetate, both in moulding compounds and in clear film (see Ch. 7, 7.4.2). A relatively small related application is as a dye carrier in the disperse dyeing of Tricel (cellulose triacetate) fibre. Here its action is essentially one of plasticising the polymer surface to facilitate penetration by a non reactive dye.

The second largest use DEP is quite unrelated to plastics technology. It is widely employed by the fragrance industry as a carrier solvent. Consistency of low odour is a key requirement and special grades are marketed for this purpose. DEP is also used as a denaturant for ethanol in some European countries.

4.5 THE PHTHALATES C_3 to C_6

Manufacturer references (see Appendix 1): 3, 4, 6, 8, 13, 14, 25, 31, 39, 40

4.5.1 C_3 PHTHALATES

Perhaps surprisingly dipropyl phthalate does not appear ever to have been used on a commercial scale.

4.5.2 C_4 PHTHALATES

di n-butyl phthalate (DBP)

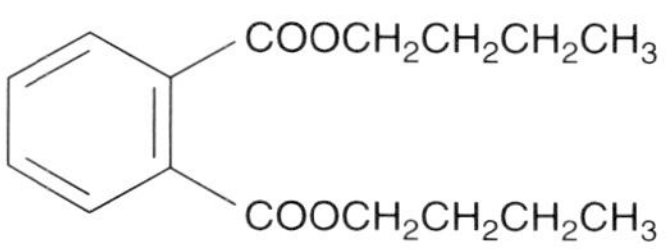

$COOCH_2CH_2CH_2CH_3$

$COOCH_2CH_2CH_2CH_3$

Viscosity at 20°C 21 mPa.s
(c.f. DOP 78)
Vapour pressure at 100°C
0.082 mbar (c.f. DOP 0.001)

Di-isobutyl phthalate (DIBP)

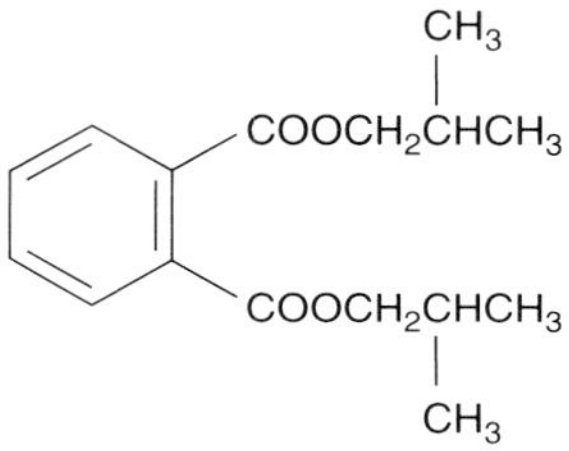

CH_3
$COOCH_2CHCH_3$

$COOCH_2CHCH_3$
CH_3

Viscosity at 20°C 39 mPa.s
(c.f. DOP 78)
Vapour pressure at 100°C
0.124 mbar (c.f. DOP 0.001)

DBP and DIBP are produced from n-butanol and isobutanol respectively, both being co-products in the manufacture of 2-ethylhexanol (see 4.6.1). Older alcohol plants using cobalt based catalysts generated a relatively high proportion of isobutanol in the product mix. Weakness of demand for isobutanol resulted in historical price levels being much lower than for n-butanol.

Consequently DIBP tended to have the status of a cheaper (and for most purposes somewhat inferior) alternative to DBP. However older plants have largely been replaced in Western Europe by newer ones using rhodium catalyst technology. This gives a much smaller proportion of isobutanol and as a consequence the price differential between the two phthalates has been greatly reduced.

DBP has been used industrially for longer than any other phthalate and has the most diverse range of applications. Whilst it was largely super-seded in PVC by less volatile longer chain phthalates in the nineteen forties it remains an important commodity product. Consumption of DBP and DIBP in Western Europe in 1992 was about 65 kte the largest application being plasticisation of PVA emulsion adhesives. DBP has had the status of being the standard general purpose plasticiser for this application (see Ch. 7, 7.2.2).

Despite the high volatility of the C_4 phthalates their use in PVC remains significant where volatile loss in processing or from the end product is not an overriding consideration. They are normally used as the fast fusing component of a mixed plasticiser system in which their impact on volatility is diluted. Comparative performance data for DBP and DIBP are included in Appendix 1. They have the highest interaction with PVC of any of the dialkyl phthalates and contribute low process temperatures, fast fusion and high extender tolerance at (historically) low cost. DIBP has higher vapour pressure than DBP and gives higher losses in volatility tests. In addition its more branched structure gives poorer cold flex and lower softening efficiency. However, as discussed in Chapter 2, 2.6.4 DIBP confers significantly better plastisol viscosity stability and hence it is preferred for some plastisol processes. The use of C_4 phthalates in PVC received some adverse publicity in the early nineteen eighties[70] when they were discovered to have caused damage to growth of certain horticultural crops in greenhouses. DIBP had been used in this case as a large part of the plasticiser system in the extruded glazing seals. Under conditions of low ventilation there had evidently been sufficient build up of vapour pressure to cause phytotoxic effects. Research indicated[71] that very small concentrations of DBP could produce these effects with specific hybrid varieties of crops. Subsequently the British Plastics Federation issued a recommendation that PVC for such application should be plasticised with less volatile plasticisers such as C_{8+} phthalates. The effect was recognised in a 1985 amendment of the British Standard for garden hose which excluded the use of C_4 phthalates. These instances of phytotoxicity were not generally thought to have any wider implications outside the sphere of horticulture.

During 1994 the EC Directives covering 'Dangerous Substances' and 'Dangerous Preparations' began to have an adverse influence on the acceptability of DBP for some major applications leading to substitution by other plasticisers. The background and implications of this are discussed in detail in Chapter 9, 9.4.

4.5.3 C_5 PHTHALATES

Until recently these have been of limited commercial significance. 'Di isoamyl phthalate' made from C_5 alcohol occurring as a byproduct from fermentation production of ethanol, has found small uses in non plasticiser applications.

Di-isopentyl phthalate produced from a synthetic alcohol feedstock is now being offered as a less volatile alternative to C_4 phthalates for plasticiser use. One application is as a fast fusing component of plastisol formulations for manufacture of cushion vinyl flooring.

4.5.4 C$_6$ PHTHALATES

n-Hexanol occurs as a component of mixed linear alcohols produced in the Ziegler process (see 4.7.1). Despite the fact that a C$_6$ phthalate could offer an valuable combination of fusion, volatility and fusion characteristics such a product has never been used in significant quantities.

'Butyl octyl phthalate' (average C$_6$) has in the past been marketed both in Europe and the USA as an inexpensive fast fusing plasticiser for PVC. It was produced by co-esterification of n-butanol and 2-ethylhexanol giving a mixture comprising DBP and DOP in addition to the butyl octyl phthalate itself. Its technical limitation was high volatility resulting from the DBP content.

4.6 BRANCHED CHAIN PHTHALATES – C$_7$ TO C$_{13}$

This group includes the large tonnage general purpose plasticisers which are phthalates of alcohols between C$_8$ and C$_{10}$. These materials are technically interchangeable in many of their applications and fierce competition usually keeps their price the lowest of all the phthalates. C$_7$ and C$_{11}$ to C$_{13}$ phthalates are used in smaller volumes for more specialised applications.

4.6.1 DOP

Manufacturer references (see Appendix 1): 3, 4, 6, 7, 8, 13, 22, 25, 30, 31, 37, 38, 39, 40, 43

Of all PVC plasticisers the one true commodity worldwide is DOP. Even after the withdrawal of smaller producers lacking feedstock integration in the last 25 years there are still more than a dozen producers spread across Western Europe. Not only is DOP the recognised standard against which other plasticisers are assessed technically but its price is a reference point and a matter of continuous concern to the whole of the flexible PVC processing industry. Figure 4.2 shows how average West European prices of suspension PVC and DOP varied between 1986 and 1994. Both commodities tend to follow the same cyclic economic trends although during 1994 and early 1995 specific raw material factors caused an unprecedented escalation in DOP price which placed it significantly above PVC. At the time of writing it is uncertain how long this situation will persist. Over the period covered by the graph the average ratio of DOP price to suspension PVC price was remarkably close to unity (1.02).

Since prices for suspension PVC and DOP tend to be similar, the raw

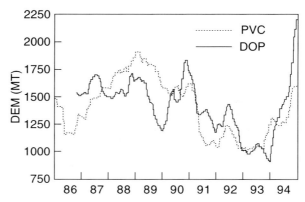

Fig. 4.2 W. European Average Prices for S-PVC and DOP 1986–94
(Source: ICIS-LOR)

materials cost of PVC compounds is (under 'normal' market conditions) little influenced by the DOP content when calculated on a weight basis. However since PVC is much denser than DOP (1.4 vs 1.0 kgl^{-1}), the **volume** cost decreases with increasing DOP level. This point is discussed in more detail in Ch. 6, 6.1.

DOP is almost unique among the phthalate plasticisers for PVC in having a single chemical component, **di-2-ethylhexyl phthalate**.

'DOP' (dioctyl phthalate) is the universal industrial designation for this material and is recognised as such in ISO 1043 Part 3 which standardises symbols for many commonly used plasticisers. This standard generally recognises accepted practice on this point rather than trying to impose much logic on the rather haphazard set of abbreviations which have evolved for industrial plasticisers over a long period of time. The growth of reported toxicological and environmental studies in recent years has given prominence to the more scientific abbreviation DEHP leading to confusion in some quarters. However both designations must continue to coexist.

The process for making 2-ethylhexanol (2-EH) starts with propylene which is a primary product from petrochemical crackers. n-Butanol and isobutanol are optional co-products from the 2-EH process.

Step 1: Hydroformylation ('oxo' reaction) of propylene

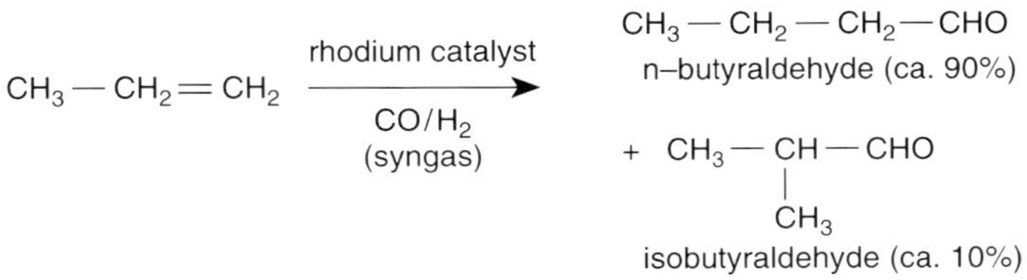

$$CH_3-CH_2=CH_2 \xrightarrow[\substack{CO/H_2 \\ (syngas)}]{rhodium\ catalyst}$$

$CH_3-CH_2-CH_2-CHO$
n–butyraldehyde (ca. 90%)

$+ \quad CH_3-\underset{\underset{CH_3}{|}}{CH}-CHO$

isobutyraldehyde (ca. 10%)

As discussed earlier, older processes using cobalt based catalysis give a higher proportion of isobutyraldehyde, in excess of 20 %.

Step 2(a): Hydrogenation of aldehydes to C_4 alcohols

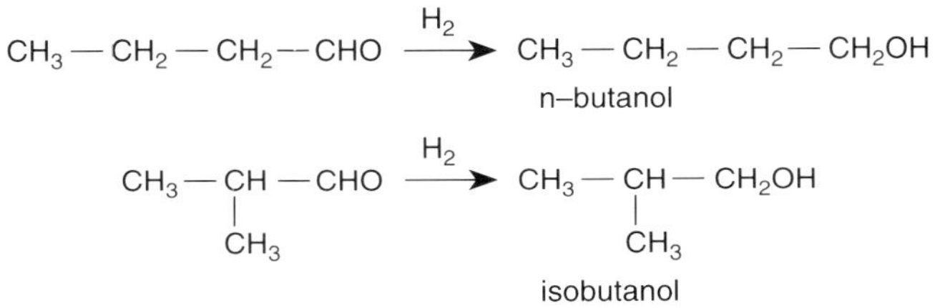

$$CH_3-CH_2-CH_2-CHO \xrightarrow{H_2} CH_3-CH_2-CH_2-CH_2OH$$
n–butanol

$$CH_3-\underset{\underset{CH_3}{|}}{CH}-CHO \xrightarrow{H_2} CH_3-\underset{\underset{CH_3}{|}}{CH}-CH_2OH$$
isobutanol

Step 2(b): Self condensation of n-butyraldehyde

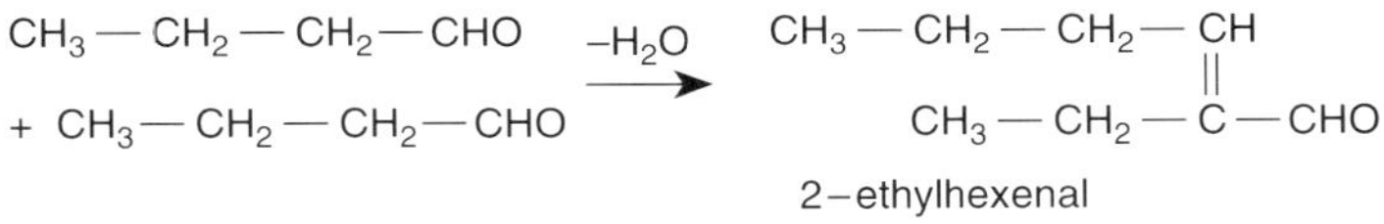

$CH_3-CH_2-CH_2-CHO$

$+ \ CH_3-CH_2-CH_2-CHO$ $\xrightarrow{-H_2O}$

$CH_3-CH_2-CH_2-\underset{\overset{||}{\underset{CH_3-CH_2-C-CHO}{}}}{CH}$

2–ethylhexenal

Step 3: Hydrogenation to 2-EH

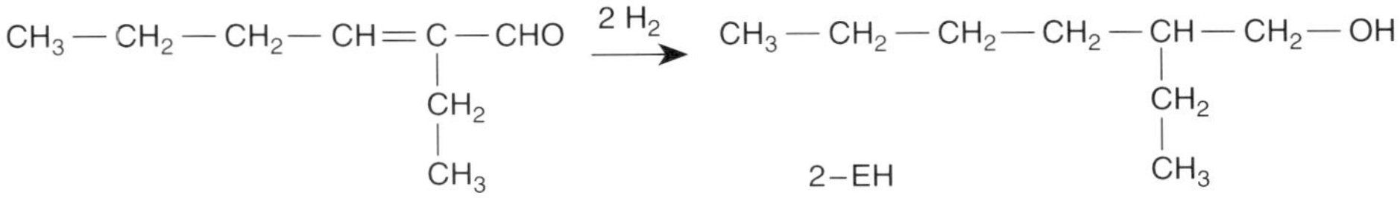

$CH_3-CH_2-CH_2-CH=\underset{\underset{\underset{CH_3}{|}}{\underset{CH_2}{|}}}{C}-CHO \xrightarrow{2\ H_2}$

$CH_3-CH_2-CH_2-CH_2-\underset{\underset{\underset{CH_3}{|}}{\underset{CH_2}{|}}}{CH}-CH_2-OH$

2–EH

Most 2-EH producers are integrated downstream into manufacture of DOP, its largest derivative. Of the three production elements of a fully integrated DOP business 2-EH represents a much larger investment than either phthalic anhydride or DOP production. Following DOP the next largest use of 2-EH is in the manufacture of 2-ethylhexyl acrylate, a raw material for coatings and adhesives.

DOP is used in nearly all applications of plasticised PVC. Constraints on its use in particularly demanding environments are imposed by considerations of volatility, extraction or migration.

In most applications DOP is subject to competition from di-isononyl phthalate and di-isodecyl phthalate which in general terms have some-what higher permanence. However DOP has some compensating advantages (commercial and technical) which enable it to retain the largest market share. Whilst it is not one of the larger applications of DOP a significant quantity is used in PVC blood bags and other medical products and it remains the only plasticiser mentioned in the relevant European Pharmacopoeia.

During the nineteen eighties there was a period of concern over the toxicological risks in using DOP following the reporting in 1980 of animal feeding tests which indicated carcinogenic potential. DOP had been one of a very limited number of materials selected for these studies on account of its extensive use and chemical simplicity. Subsequent studies placed these results in proper perspective as discussed in Chapter 9 and provided sufficient reassurance to make restrictions on the use of DOP unnecessary. However, in the short term there was some tendency to substitute DOP by technical alternatives. This was essentially a DOP avoidance exercise in response to pressures from certain downstream customers since there was no evidence to indicate greater safety for the substitute plasticisers, none of which had been tested to the same extent as DOP. In Europe there are some examples of DOP exclusion dating from this period.

As with dibutyl phthalate, the EC Dangerous Substances Directive has led to DOP being classified by suppliers as – Toxic to Reproduction, Category 3. 'Preparations' containing 5% or more of DOP are also classified in this way (in the context of DOP this applies mainly to plastisols). Containers holding DOP or DOP preparations must now carry warning labels which **appear** to indicate that its use involves a higher level of risk than alternative plasticisers. However, in contrast to the DBP market, substitution of DOP as a result of this factor has so far been of minor significance. The background to this situation is discussed in detail in Chapter 9 (9.3 and 9.4).

4.6.2 POLYGAS OLEFIN DERIVATIVES (THE 'ISO-ALCOHOL' PHTHALATES)

The following manufacturer references (see Appendix 1) also cover the products described in the following section (4.6.3): 3, 4, 6, 13, 22, 25, 29, 30, 31, 37, 40

These are a group of phthalates ranging from C_7 to C_{13} of which the most important members are di-isononyl phthalate and di-isodecyl phtha-late. The feedstock alcohols are produced by hydroformylation of olefins from large 'polygas' plants attached to refineries. These units oligomerise

C_3 and C_4 olefin mixtures to produce hydrocarbons suitable for use in gasoline. They are capable of producing olefins ranging from C_6 to C_{12} which can be separated by distillation into various feedstocks for plasticiser alcohols.

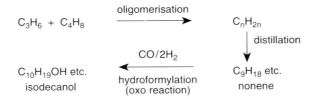

The products from this sequence have primary alcohol structures with a fairly high degree of methyl branching. The range of alcohols obtained is indicated in Table 4.3.

TABLE 4.3
PLASTICISER ALCOHOLS FROM POLYGAS OLEFINS.

ALCOHOL	Approx. methyls per chain	Olefin origin
C_7 isoheptanol	2	C3 + C3
C_8 iso-octanol	2	C3 + C4
C_9 isononanol	2	C4 + C4
C_{10} isodecanol	3	C3 + C3 + C3
C_{11} isoundecanol	3	C3 + C3 + C4
C_{12} isododecanol	3–4	C3 + C4 + C4
C_{13} isotridecanol	4	C3 + C3 + C3 + C3 and C4 + C4 + C4

The dominant global producer of alcohols of this type is Exxon Chemical. Uniquely, Exxon exploits the full range of polygas combination shown above in a range of phthalates with a wide spectrum of performance. Graphs constructed from Exxon's published data[72] provide an instructive illustration of the effect of alkyl chain size on physical properties and plasticising performance for a series of closely related phthalates. These are shown in Figures 4.3–4.5.

DIHP
Di-isoheptyl phthalate is not included in the largest tonnage category of general purpose phthalates. It finds application where its slightly lower viscosity and faster fusion characteristics give some advantage. It has established a strong position in formulations for cushion vinyl flooring. Here its use allows a reduction in the level of (historically) relatively

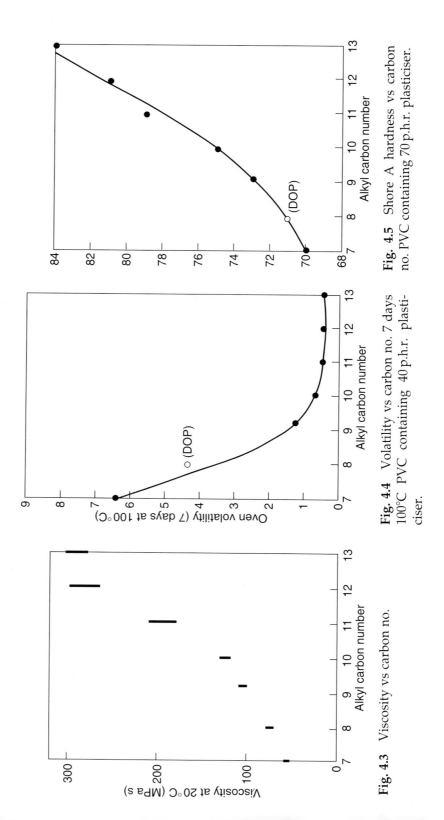

Fig. 4.5 Shore A hardness vs carbon no. PVC containing 70 p.h.r. plasticiser.

Fig. 4.4 Volatility vs carbon no. 7 days 100°C PVC containing 40 p.h.r. plasticiser.

Fig. 4.3 Viscosity vs carbon no.

expensive butyl benzyl phthalate. The main technical impediment to its wider use is its volatility which is somewhat higher than that of DOP.

DIOP

Historically di-isooctyl phthalate has been an important general purpose plasticiser and was available for very many years as a close alternative to DOP. Whilst the two have always been regarded as directly inter-changeable for most purposes their structural difference does give differ-ences in some aspects of performance. DIOP has lower thermo-oxidative stability and more often requires the use of an antioxidant. It confers rather poorer cold flex than DOP. Surprisingly its volatility from PVC is somewhat lower than for DOP.

   The use of DIOP declined concurrently with the growth of DINP and it has recently disappeared from the European market, alternative uses having been found for the heptene precursor of iso-octanol.

DINP

Polgas-derived di-isononyl phthalate appeared in the market much more recently than DIOP and rapidly achieved a large sales through a combina-tion of differentiated performance and competitive pricing (versus DOP). It also provided a technically convenient option for users preferring to avoid the use of DOP. In general terms its performance is half way between DOP and DIDP and it is close enough to DOP to be a close competitor in all but a few applications. Its lower volatility has been a factor in gaining market share. One area where it has become the predominant plasticiser is PVC wallcoverings.

   The polygas-derived isononanol used to manufacture DINP does not consist entirely of C_9 alcohols but has a C_{10} content of about 20%. Other types of 'isononanol' from different sources are discussed in the next section.

DIDP

Di-isodecyl phthalate has been a major plasticiser for far longer than DINP. Whilst it is interchangeable with DOP for many applications this requires a greater adjustment of formulations and processing conditions than in the case of DOP-DINP interchange. DIDP has lower plasticising efficiency than DOP necessitating higher levels of addition to PVC. Depending on the relative prices of phthalates and PVC at any time this can lead to lower compound costs despite the fact that DIDP is generally slightly more expensive than DOP.

   There are large well established applications of DIDP where DOP is too volatile to be acceptable. Conversely, the slower fusion rate of DIDP and its requirement for higher process temperature make it less attractive for some application where the use of DOP is standard. The largest market for DIDP is cable covering compounds and it has the status of being the main plasticiser used in Europe in these compositions. Because of the low volatility of DIDP these are capable of meeting a wider range of specifications than when DOP is used. Even so, for a significant proportion of cable applications DOP is quite acceptable and selection rests simply on cost differentials.

Because DIDP is susceptible to oxidation at higher service temperatures, the advantage in heat ageing resistance resulting from its low volatility is only realised in the presence of an antioxidant. Consequently a 'stabilised' grade (typically containing 0.5% of Bisphenol A) tends to be the standard form in which it is available. The user may need to exclude the antioxidant for reasons of (e.g.) lack of food contact approval or to avoid colour formation through interaction with blowing agents. If so, it may be necessary to specify this to the supplier.

DIUP

Compared with DIDP, di-iso-undecyl phthalate is a step further down the line towards low volatility at the expense of ease of processing. It therefore lacks the balance of performance characteristics required of a general purpose plasticiser. Whilst it is somewhat more expensive than these commodity products it presents a relatively low cost option for meeting various specifications requiring low volatility. It has been used in mixtures with trimellitates as a means of reducing the cost of high temperature cable formulations. As with DIDP it is supplied containing antioxidant. The performance of DIUP does not match that of linear undecyl phthalate (DUP) which confers both better heat ageing resistance and better low temperature properties.

Jayflex UDP

This plasticiser, described as undecyl dodecyl phthalate is produced from mixed alcohols with an average carbon number of approximately 12 (as indicated by the manufacturer's published saponification number). Its performance is generally similar to that of DIUP.

DITDP

Antioxidised di-isotridecyl phthalate, alternatively known as ditridecyl phthalate, DTDP, has a long history of use as a high temperature plasticiser for PVC cable compounds. Its high PVC solution temperature (160°C, c.f. DOP 124°C, DIDP 140°C) places it on the useful limit of PVC compatibility and processability. It has the highest molecular weight of all the phthalate plasticisers. Its softening efficiency is much lower than that of the general purpose phthalates although it confers similar cold flex. It has outstanding resistance to extraction by soapy water and similar media.

During the nineteen seventies DITDP was largely displaced by trimellitates giving better processability and even lower volatility. However it continues to maintain a place in the cable industry in applications where its price/performance is advantageous.

4.6.3 OTHER C$_9$ AND C$_{10}$ PHTHALATES

In addition to the main C$_9$ phthalate, the polygas-derived DINP described above, there are two other types having different degrees of branching and correspondingly differing performance. One version, di-3,5,5-trimethylhexyl phthalate has been on the market for more than thirty years. The highly branched alcohol feedstock is produced by hydroformylation of di-isobutene. In view of its predating of DINP this product has, confusingly been known as dinonyl phthalate, DNP. Its structure gives a performance which in some respects is poorer than for less branched C9 phthalates, in particular worse cold flex and slightly higher volatility. However in other ways this product shows remarkably good performance being very resistant to both oxidation and hydrolysis. These can be attributed to steric shielding by the doubly branched oxidation resistant 5 position in the carbon chain.

Commercially more significant is the third type of C9 phthalate which has lower branching than the standard polygas derived product. Dimerisation of an n-butene feedstock (the 'Octol' process) gives a C8 olefin which is essentially monomethyl heptene. Hydroformylation then produces a mixture of C9 alcohols which are singly (the major part) or doubly branched. The phthalate produced from this feedstock has higher performance than standard DINP, specific advantages being lower volatility and better cold flex. Its perceived cost/performance benefits have given it a significant place in the market, partly at the expense of linear phthalates. Although it has sometimes been described as 'semilinear' a more accurate term for this phthalate is 'lightly branched'.

There is currently some interest in a new type of C10 phthalate[73] which could offer performance advantages relative to the old established DIDP. Technology exists for using n-butene in an alcohol process which parallels the manufacture of 2-ethylhexanol from propylene (see 4.6.1). The product is 2-propylheptanol with n-pentanol as an optional co-product. In the absence of comparative performance data it can be speculated that di-2-propylheptyl phthalate could achieve a position analogous to the lightly branched C_9 phthalate described above. Its future status will rest on the market's perception of its cost/performance relative to established phthalates.

CH$_2$——CH$_2$——CH$_3$ ... (structure)

di–2–propylheptyl phthalate

4.7   LINEAR PHTHALATES

Manufacturer references (see Appendix 1): 3, 4, 6, 8, 13, 15, 31, 37

4.7.1   FEEDSTOCKS

The plasticisers marketed as 'linear' phthalates are produced from alcohol mixtures containing anything between 50 and 100% of straight chain isomers, the non-linear components having a low degree of branching. In contrast to the branched phthalates described earlier in this chapter most linear phthalates are based on alcohol feedstocks with a spread of chain lengths which can be quite wide, for example C_7 to C_{11}. Linear alcohols of suitable chain length are nearly all derived from ethylene although limited quantities linear alcohols obtained from natural products have also been used. Most linear alcohols for plasticisers are co-products from the production of very much larger quantities of longer chain alcohols as detergent feedstocks. There is some overlap of chain length between the alcohols used in these two different markets. For example Linevol® 911 and Dobanol® 911 produced by Shell Chemical have essentially the same composition, the former being well known as a plasticiser feedstock and the latter being the lowest of a range of detergent alcohols.

Starting from ethylene there are two basic routes to linear alcohols. The longer established one uses the Ziegler process which adds sequences of ethylene molecules to triethyl aluminium giving as an intermediate trialkyl aluminium with a range of chain lengths. Oxidation and hydrolysis removes the oligermerised ethylene as a mixture of 100% linear alcohols having **even** numbers of carbon atoms, n-C_4H_9OH, n-$C_6H_{13}OH$, n-$C_8H_{17}OH$ etc.

$$Al(CH_2-CH_3)_3 + n\ CH_2=CH_2 \longrightarrow Al\left[(CH_2-CH_2)_{n+1}H\right]_3$$

$$\downarrow O_2$$

$$Al(OH)_3 \atop + 3\ CH_3CH_2(CH_2-CH_2)_nOH \xleftarrow{\ H_2O\ } Al\left[O(CH_2-CH_2)_{n+1}H\right]_3$$

The mixed product is then fractionated to give cuts suitable for detergent or plasticiser feedstocks.

The other type of process first oligomerises ethylene to produce linear olefin mixtures which are fractionated to give feedstocks for production of alcohols or other derivatives. The olefins are converted to alcohols by hydroformylation which adds one carbon atom to the chain. Thus a C_n olefin is converted to a C_{n+1} alcohol.

$$\text{e.g.} \quad n-C_9H_{18} \xrightarrow[\text{catalyst}]{CO/H_2} n-C_9H_{19}CHO \xrightarrow{H_2} n-C_{10}H_{21}OH$$
$$\text{n-nonene} \qquad\qquad\qquad\qquad\qquad \text{n-decanol}$$

Alcohols obtained by this route would be expected to contain only components with **odd** numbers of carbon atoms. The exception to this is the range of products derived from the Shell Higher Olefin Process (known as SHOP). This involves a metathesis step by which mixtures containing unwanted short and long chain olefins react to give mixtures within the required range containing a proportion of **odd**-numbered components. Hence the Linevol® alcohols derived from SHOP olefins have both odd and even numbered chain lengths.

The conversion of linear olefins to alcohols results in some loss of linearity depending on the characteristics of the hydroformylation catalyst. The least branching is found in the Shell alcohols which typically comprise about 83% straight chain isomers. Other alcohol producers achieve linear contents around 60 or 70% as indicated in the following review of individual products. It should be realised that the branching of the non-linear portion of these products is of a low order and consequent differences in performance are considered insignificant by most users. The concept of phthalate linearity and its consequences is treated in more detail under 4.8.

Esterification of phthalic anhydride with mixed alcohol gives a phthalate mixture in which the ratio of the components is determined statistically by the composition of the feedstock. An equimolar mixture of ROH and R'OH will give a mixed phthalate with three components, RR, RR' and R'R' in the molar ratio 1:2:1 (provided the alcohols have equal reactivity). Starting with the two alcohols in the molar ratio of 1:9 gives the phthalates in the ratio 1:18:81. Three alcohol components give a phthalate with six components, four give ten and so on.

As already indicated technical mixtures of this type have the same plasticising performance as a single component of the same average molecular weight and linearity. However when the components have a large spread of molecular weight this correspondence can break down for some aspects of performance. In particular, lower molecular weight components can have adverse effect on volatility and windscreen fogging.

4.7.2 COMMERCIAL LINEAR PHTHALATES

These are now reviewed in order of average chain length.

C6
n-Hexanol is available as a specific cut from the alcohols produced by the Ziegler process. The chain length is too low for the main marketable advantages of linear phthalates to be demonstrated and di-n-hexyl phthalate has so far been little used.

C7
n-Heptanol is available as a byproduct from splitting of castor oil but its price and availability are variable. The phthalate has been produced commercially and used in plastisol coated products where it has provided advantages in process speed and cold flex relative to DOP without any disadvantage in volatility.

C8
C6/C8/C10 alcohol fractions from the Ziegler process have been well known phthalate feedstocks for more than 30 years. The phthalates compete technically with the 80% linear C7/C8/C9 phthalate derived from Shell alcohols. The somewhat lower average molecular weight of the latter combined with the difference in linearity gives it slightly higher volatility and poorer cold flex. These phthalates are used in coated fabrics etc. for outdoor service and in specialised low temperature cables.

Perhaps surprisingly, di-n-octyl phthalate has never been of commercial significance. However process technology exists in Japan for production of n-octanol from butene and the phthalate has established a small position in the Japanese market. If it were to be introduced in Europe it would be expected to compete in the first instance with the established C8 and C9 (average) mixed linear phthalates reviewed here.

C9

Consumption of linear phthalates in the USA is on a much larger scale than in Europe. Much of the volume is accounted for by a single product in the C9 category. This is the C7/C9/C11 phthalate introduced in 1970 by Monsanto as Santicizer 711 which is now supplied by BASF as Palatinol 711. The difference in its market status between the USA and Europe is mainly due to differences in price structures. In Europe where its consumption is relatively small this product tends to be regarded as a lower cost/lower performance alternative to linear 911 phthalates. A linear 7–11 phthalate based on Shell alcohols is also available.

A 100% linear C8/C10 phthalate from a Ziegler alcohol fraction is one of the longest established of all linear phthalates. A slightly different product from a fraction with a lower C8 content is now marketed and this competes more directly with the C9–C11 phthalates. 100% linear C8/C10 alcohols occurring as byproducts from the manufacture of detergent intermediates from natural products have also been used as phthalate feedstocks.

C10

Products with average chain length close to C10 constitute the main group of linear phthalates used in Europe, the major proportion being based on C9–C11 alcohols. These are either 80% straight chain C9/C10/C11 mixtures (Shell Chemicals Linevol® 911) or C9/C11 mixtures containing about 60% of straight chain isomers. These are marketed as speciality plasticisers with prices which have historically been double those of commodity phthalates. Their principal applications are nonfogging automotive interior trim and long life outdoor products such as single ply roofing membranes.

A further phthalate competing with this group of products and slightly superior in some aspects of performance is based on the highest of the plasticiser feedstocks obtained from the Ziegler process. This is a C10/C12 fraction with average chain length approximately 10.4.

C11

Diundecyl phthalate (DUP) represents the upper useful limit of chain length for linear phthalate plasticisers. Products on the market are manufactured from alcohols with C11 content close to 100% containing between 50 and 70% of straight chain isomers. Higher chain lengths and/or higher linearity would tend to give problems of high freezing point and limited PVC compatibility. DUP is usually supplied in an antioxidised grade for use in cable insulation. It is particularly useful for meeting specifications requiring resistance to heat ageing, cold crack and hot deformation.

C17

With chain lengths above C11 linear phthalates become solids at ambient temperature and their PVC compatibility is limited. Cetyl stearyl phthalate produced from a natural product-derived C16/C18 alcohol mixture is used in significant quantities in rigid PVC as a processing lubricant, mainly in the extrusion of profiles for window frames.

4.8 DIALKYL PHTHALATE PLASTICISERS FOR PVC: STRUCTURE, PERFORMANCE AND COST

4.8.1 BRANCHING INDEX

In order to make comparisons between the full range of linear and branched phthalates on a consistent basis it is necessary to select a common scale of linearity/branching covering everything from 100% linear (0% branched components) to 'highly branched'. The extreme example of the latter is di-3,5,5,-trimethylhexyl phthalate ('dinonyl phthalate'). The scale of '%linearity' (meaning percentage of straight chain components) commonly applied to linear phthalates is clearly of no use for differentiating between the various branched chain phthalates with different degrees of branching. The index used in the following discussion, which can describe any phthalate, is the percentage of the total carbon atoms of the alkyl groups which is contained in the side chains. Hence for a C_9 phthalate consisting of dimethyl hexanols the branching index would be 22%. Since most branching in plasticiser alcohols consists of methyl groups, the complication of longer branches can be largely disregarded. A rare exception is illustrated in a comparison of DOP and DIOP both of which have a branching index of 25%.

DOP alkyl group $-CH_2-CH-CH_2-CH_2-CH_2-CH_3$
                          |
                   CH_2-CH_3

Typical DIOP alkyl group $-CH_2-CH_2-CH-CH_2-CH-CH_3$
                                      | |
                                    CH_3 CH_3

Of the two, DOP generally shows the more 'linear' performance characteristics. Cold flex and softening efficiency are better than for DIOP presumably because the ethyl side group provides an additional contribution to the plasticising mechanism. Oxidative stability is very much greater for DOP because not only does its structure contain less branching points which are the sites for oxidation, but they are also more shielded sterically than in DIOP. Di-2-isopropylheptyl phthalate (see 4.6.3) would be expected to show the same advantages relative to DIDP.

4.8.2   EFFECT OF CARBON NUMBER AND BRANCHING ON PROPERTIES AND PERFORMANCE

Technical or economic advantages as PVC plasticisers are indicated by (+) and disadvantages by (−).

Increasing carbon number results in:
- falling density, giving a small advantage in volume cost (+)
- increasing viscosity, this being most marked for branched phthalates (−)
- increased melting point for the linear phthalates, causing handling and compatibility problems at high chain lengths (−)
- decreasing vapour pressure, hence lower volatile loss in processing and service (+)
- higher PVC solution temperature (+/−)
- reduced compatibility and lower extender tolerance (−)
- the need for higher process temperatures for dryblending and plastisol fusion (−)
- better plastisol viscosity stability (+)
- lower softening efficiency (+/−) depending on weight and volume costs relative to PVC
- improved cold flex (+)
- slower migration (+)
- reduced marring of polar polymers in contact with PVC (+)
- a tendency to greater extraction by hydrocarbon oils (−)

Increased branching results in:
- increasing viscosity (−)
- a tendency to supercool to a glass rather than form a crystalline solid (+)
- increasing vapour pressure, hence higher volatility in processing and service (−)
- somewhat lower softening efficiency (+/−)
- markedly worse cold flex (−)
- slightly higher PVC compatibility and extender tolerance (+)
- higher plastisol viscosity and greater dilatancy (−)

- slower migration (+)
- lower thermo-oxidative stability (−)
- lower photostability (−)
- greater resistance to chemical hydrolysis (+)
- greater resistance to biodegradation (+)

4.8.3 PHTHALATE VALUE VERSUS STRUCTURE

Figure 4.6 places the various phthalates reviewed in Sections 4.6 and 4.7 on a carbon number/branching matrix, the branching scale being the index defined in 4.8.1. Since there is a lack of published information on the isomeric content of some of the branched phthalates the branching indices shown for these products are estimates which may be subject to some error. However this will not significantly detract from the main points arising from this treatment. The phthalates are identified in the diagram as follows, those made from alcohols with a spread of carbon number being shown as rectangles, products of a single carbon number as circles.

BRANCHED	DIHP	di-isoheptyl phthalate
	DIOP	di-iso-octyl phthalate
	DINP (1)	di-isononyl phthalate (polygas type)
	DINP (2)	di-isononyl phthalate (Octol type)
	DNP	di-3,5,5,-trimethylhexyl phthalate
	DIDP	di-isodecyl phthalate
	DIUP	di-iso-undecyl phthalate
	UDP	(Jayflex UDP)
	DITDP	di-isotridecyl phthalate
LINEAR	DnHP	di-n-heptyl phthalate
	DnOP	di-n-octyl phthalate
	610P	C6/C8/C10 phthalate of Ziegler alcohols
	79P	C7/C8/C9 phthalate of Shell alcohols
	812P	C8/C10 (minor C12) phthalate of Ziegler alcohols
	711P	C7–C11 phthalate (two similar versions)
	911P	C9–C11 phthalate (two similar versions)
	1012P	C10/C12 phthalate of Ziegler alcohols
	DUP	di-undecyl phthalate (two similar versions)

In figure 4.6 the distinction between linear and branched phthalates as separate groups becomes somewhat blurred. Mutual proximity of different materials on the diagram indicates ease of technical interchangeability unless they happen to fall on different sides of a strict borderline imposed by some end use specification.

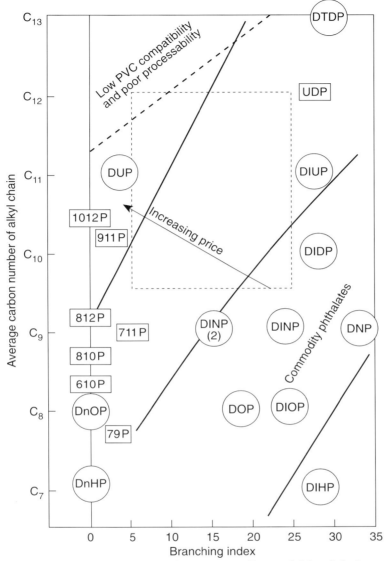

Fig. 4.6 Structure classification of commercially available phthalates.

Outside the small group of high tonnage general purpose plasticisers, the price of phthalates tends to increase with decreasing branching and to a lesser extent with increasing carbon number. The most expensive materials are found in the top left hand corner of the diagram, i.e. the most linear phthalates with average chain length C10 to C11. The diagonal lines superimposed on the grid are intended to show the historical (ca. 1990) pattern of relative prices rather than actual values. In summary, the market places high value on the technical benefits conferred by low branching and high carbon number (up to the limits imposed by processing and compatibility considerations).

An interesting feature of the diagram is the empty area bounded by the limits: carbon number C9.5 to C12, branching index 5% to 25%. This results from a historical lack of feedstocks in this area. Bearing in mind the progressive increase in interest in improved plasticiser permanence there would appear to be opportunities for phthalates of this type if suitable feedstocks became available. A specific example would be di-2-propylheptyl phthalate (see 4.6.3). Like DOP its branching index is arbitrarily reduced to allow for the fact that it is not multi-methyl branched. Clearly the smaller the price premium over commodity phthalates (i.e. the greater the distortion of the price contours in Figure 4.6) the greater would be the potential consumption of any product appearing in this empty quarter.

4.9   OTHER PHTHALATES

4.9.1   Butyl benzyl phthalate (BBP)

Manufacturer references (see Appendix 1): 3, 5, 35

BBP is known by PVC technologists as the industry's standard fast fusing plasticiser. It continues to be used in significant quantities in this function. BBP can be produced commercially by conventional esterification of phthalic anhydride with a mixture of butanol and benzyl alcohol. However this route gives a technically inferior product containing dibutyl phthalate and dibenzyl phthalate in addition to the main component. The standard process for making BBP is more specialised and gives a single component product in three steps:

(a) monoesterification

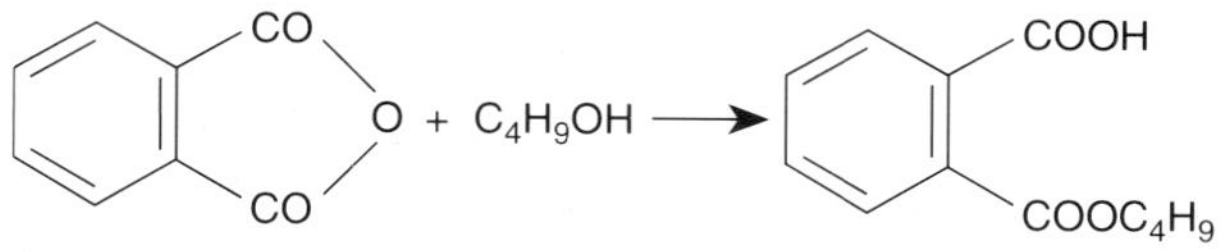

(b) neutralisation

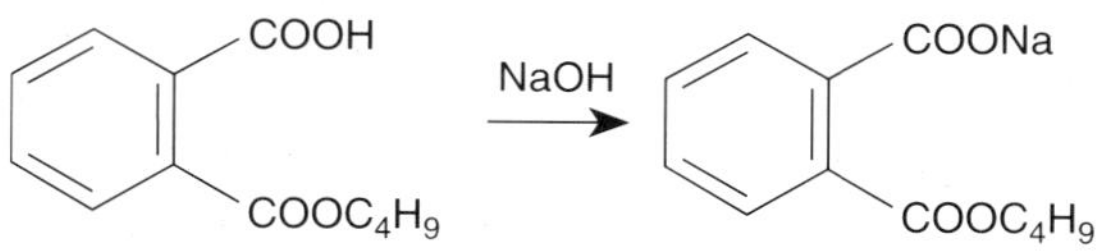

(c) reaction with benzyl chloride

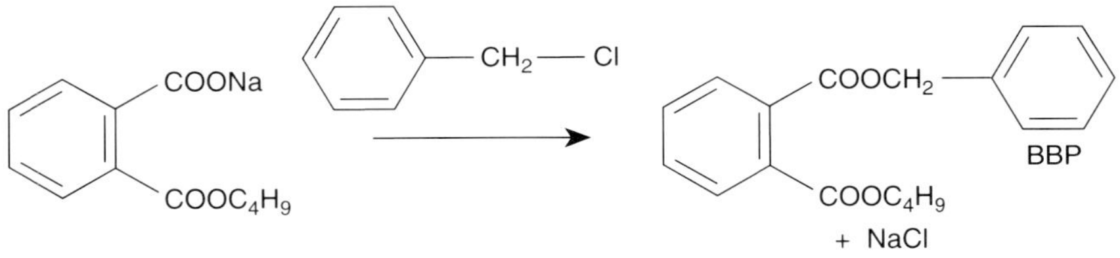

The generation of sodium chloride in step (c) necessitates the use of special corrosion resistant plant.

The specific technical attributes of BBP are due to the presence of a second polarisable benzene ring in its structure. This gives additional interaction with PVC compared with dialkyl phthalates of the same molecular weight. Solubility parameter (see Chapter 1, 1.7) is a common theoretical predictor of plasticiser/polymer compatibility, high compatibility being indicated by a close match of values. Of all the dialkyl phthalates dibutyl phthalate provides the closest match to PVC and BBP is even closer.

	Solubility parameter δ
BBP	9.88
PVC	9.66
DBP	9.41
DOP	8.43

Whilst both DBP and BBP are used as fast fusing plasticisers, BBP is much less volatile and therefore a more acceptable option in most cases. Comparative performance data are contained in Appendix 1.

The main traditional area of application of BBP is in PVC cushion vinyl flooring where its highly solvating character provides advantages of fast fusion, high expansion and good cell structure in foam layers and in wear layers relatively high resistance to staining. The enhancement of foam expansion by the use of BBP is illustrated in Table 4.4 which shows the effect of partial replacement of DOP by BBP.[74] Expandable plastisols containing an azodicarbonamide blowing agent were fused on a semi–industrial scale coating line. The coating weight was 650 g/sq m, oven length 6 metres and oven temperature 200°C.

A further performance feature of BBP is the high level of gloss and sparkle (or brilliance) conferred to PVC. The causes of this effect are believed to be twofold. Firstly its efficiency in solvating the polymer prevents the surface matting effect caused by retention of the resin's

TABLE 4.4
EFFECT OF BBP ON PVC EXPANSION

	Formulation 1	Formulation 2
PVC paste polymer	100	100
DOP	80	-
Santicizer 160 (BBP)	6	6
AZDN blowing agent (50% DOP)	6	6
Ca/Zn kicker	2	2
Pigment	2	2
Silica dispersion	3	1.5
Degree of foam expansion (%)		
- at coating rate 1.8 min^{-1}	267	358
- at coating rate 2.6 min^{-1}	110	195

particulate character. Secondly the effect on brilliance is associated with the relatively high refractive index of BBP (n_D^{20} = 1.540, c.f. DOP 1.485).

Although it confers distinct advantages, the structure of BBP leads to some performance penalties. The low proportion of mobile alkyl carbon atoms results in poor low temperature performance. As illustrated in Chapter 3, 3.10 its high aromatic content causes high UV absorbtion and consequently detracts from the photostability of PVC.

As a consequence of its high interaction with PVC BBP tends to give poor viscosity stability in plastisols. In addition its lack of linearity tends to confer dilatancy to plastisols. A number of proprietary blends based on BBP have been developed aimed mainly at the flooring industry and containing non-solvating low viscosity components to allow the formulation of spreadable plastisols with low plasticiser content. In the past these low viscosity materials have been volatile materials such as dodecyl benzene. However there has been concern in some countries over the exposure of building occupants to vapours released from installed floor-covering. This has lead to the recent development of blends incorporating much less volatile viscosity reducing components (Santicizer® 1700 series).

In addition to its familiar PVC outlets, BBP finds use as a plasticiser in sealants based on other polymers, particularly acrylics with which it has high compatibility. Here its applications overlap those of the other benzyl phthalates described in the next section.

4.9.2 OTHER BENZYL PHTHALATES

Manufacturer references (see Appendix 1): 35

The process used to manufacture BBP has the potential for producing a wide range of alkyl benzyl phthalates by replacing butanol by other alcohols in the monoesterification step. In practice two such products are commercially established. One is based on 2-ethylhexanol (octyl benzyl phthalate, Santicizer® 261). The other (Santicizer® 278) is a high molecular weight product which uses as its alcohol feedstock Texanol® (see Ch. 5, 5.11).

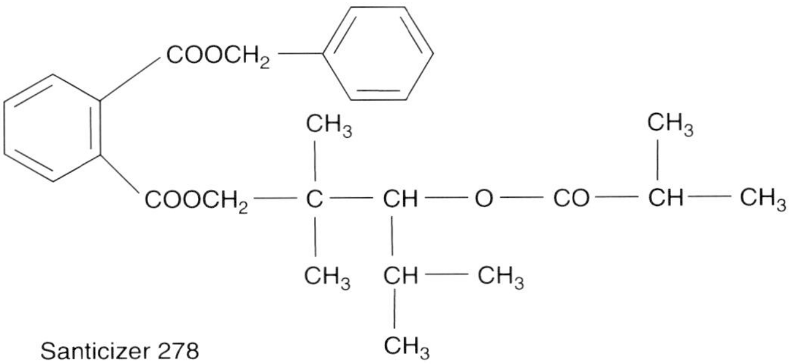

Santicizer 278

To a degree octyl benzyl phthalate shares the compatibility and processing characteristics of BBP but has the advantage of lower volatility. The high molecular weight of Santicizer 278 gives it the lowest volatility of this group of products. In addition its resistance to extraction by hydrocarbons is in the same class as polyesters over which it has the advantage of better processability. Its bulky inflexible molecular structure gives it low plasticising efficiency and poor low temperature properties. The main application of this material is as a plasticiser for polysulphide sealants for double glazing units. Here its low vapour pressure prevents internal fogging which can result from evaporation and condensation of more volatile plasticisers.

4.9.3 BISPHTHALATES

In the early nineteen sixties an adaptation of the BBP production route gave rise to an unusual family of products with valuable performance as PVC plasticisers. Reaction of one mole of a dichloride with two moles of a sodium alkyl phthalate gave a high molecular weight bisphthalate. This type of product combined some of the permanence of polyesters with the processing advantages of monomeric plasticisers. The general structure was as follows:

For a variety of reasons the bisphthalates never became established commercially and like a number of other technically interesting plasticisers they rest largely forgotten.

4.9.4 ALICYCLIC PHTHALATES

The following manufacturer references also cover the materials described in the following section (4.9.5): 5, 6, 14, 18, 24, 29

Dicyclohexyl phthalate (DCHP) is a crystalline solid with a melting point of 63°C.

It is used in conjunction with dibutyl phthalate as a plasticiser in the heat sealable nitrocellulose surface layer of cellophane film. Another use of DCHP which has now declined to a very low level is in heat activated adhesives which have been largely superseded by the hot melt type. In this application fine solid particles of DCHP were dispersed in an acrylic emulsion. This was spread on the substrate, usually paper, to form a dry, tack-free coating. When adhesion was required the temperature was raised above the melting point of DCHP resulting in plasticisation thereby giving the degree of tack necessary for bonding.

   Other alicyclic phthalates which have had small specialised uses have been di(methylcyclohexyl) phthalate and butyl cyclohexyl phthalate.

4.9.5 PHTHALATES OF GLYCOL ETHERS

The glycol ethers referred to here are primary alcohols manufactured by ethoxylation of methanol, ethanol and butanol. They are well known as solvents and as feedstocks for acetate solvents. Phthalates based on these alcohols have limited but long established uses as plasticisers for polymers other than PVC. The two best known such products are di-(methoxyethyl) phthalate and di-(butoxyethyl) phthalate.

COOCH₂CH₂OCH₃ ... COOCH₂CH₂OCH₃ —— di-(methoxyethyl) phthalate also known as di-(methylglycol) phthalate (DMGP) and di-(methyl Cellosolve®) phthalate

COOCH₂CH₂OCH₂CH₂CH₂CH₃ ... COOCH₂CH₂OCH₂CH₂CH₂CH₃ —— di-(butoxyethyl) phthalate also known as di-(butylglycol) phthalate (DBGP) and di-(butyl Cellosolve®) phthalate

The ether groups present in these structures gives them additional interaction with polar polymers where they have the same compatibility as simple dialkyl phthalates of lower molecular weight. Thus DMGP can be used as an alternative to DEP in cellulose acetate and DBGP as an alternative to DBP in PVA giving benefits of lower volatility but at higher cost.

Higher molecular weight phthalates produced from polyethoxylated alcohols have been evaluated in PVC and their use as antistatic plasticisers has been patented.[75]

CHAPTER 5

Other Plasticisers

5.1 INTRODUCTION

Although phthalates predominate in tonnage terms the great majority of plasticisers in commercial use are of other types. Collectively they account for about 15% of consumption. As with phthalates the main demand is in PVC and there is a corresponding emphasis on PVC throughout this chapter. Again DOP is used as the standard technical and commercial yardstick. In nearly all cases except the important class of chlorparaffins these plasticisers are considerably more expensive than DOP and the other general purpose phthalates (comments on relative cost/performance in this and the following chapter make the assumption that the very high prices of commodity phthalates during 1994–95 will prove to be a temporary aberration). Non-phthalate plasticisers are therefore usually confined to applications where the phthalates are deficient in some aspect of performance and many are marketed as speciality products. They are frequently used in mixed plasticiser systems which include phthalates in order to optimise cost/performance.

This chapter first examines the range of specialised performance characteristics which are used to categorise plasticisers and then goes on to review commercially available plasticisers by chemical type.

5.2 CLASSIFICATION BY FUNCTION

In marketing plasticisers suppliers often use traditional descriptions to categorise them according to the primary attributes differentiating them from general purpose products. Such labels can aid selection although they are inevitably simplistic since all plasticisers possess multiple attributes to varying degrees. The main categories are reviewed here.

'Fast fusing'
Fast fusing plasticisers are those having low PVC solution temperatures. For a given rate of heat input they achieve key stages of processing (e.g. dryblend formation, full fusion and optimum mechanical properties) after short times. They have solubility parameters closely matching that of PVC

as a result of their polar and polarisable molecular structures. Fast fusion is favoured by low molecular weight and so these materials can be relatively volatile.

The archetype of fast fusing plasticisers is butyl benzyl phthalate.

'Low viscosity'

This usually indicates low viscosity conferred to plastisols at a wide range of shear rates rather than low viscosity per se. An associated attribute is good plastisol storage stability. Plasticisers falling into this category generally do have low viscosity themselves combined with a low level of interaction with PVC, the latter being indicated by relatively high solution temperature. They are mainly aliphatic esters.

'High temperature'

High temperature plasticisers are used in end products designed to withstand high service temperatures, for example cables for continuous operation at 105°C. This requires materials with low volatility together with good thermo-oxidative stability. The latter is normally achieved by incorporation of an antioxidant, sometimes indicated in the plasticiser designation as 'stabilised'. The use of these high temperature plasticisers does not overcome the intrinsic limitation of plasticised PVC associated with thermoplastic flow. Although careful selection of plasticiser can have a beneficial effect, large reductions in hot deformation require the use of a high molecular weight polymers or the inclusion of cross-linkable additives. The most common high temperature plasticisers for PVC are trimellitates. C11 to C13 phthalates offer a somewhat lower level of performance.

'Non-fogging'

This term is of significance in the selection of plasticisers for automotive interior trim components. Whether or not a plasticiser is non-fogging depends on the severity of the target specification. An ability to meet the most stringent specifications depends on two factors. Firstly the main component or components of the plasticiser must have low vapour pressure, and must therefore be of relatively high molecular weight. In addition there is a need for tight constraints on the levels of any impurities having high vapour pressure. Hence it may be necessary for user and supplier to agree a specification for a material of 'automotive quality' (see Chapter 8). In recent years trimellitates have become established at the top of the non-fogging league.

'Low temperature'

The use of low temperature plasticisers allows the formulator to achieve a specified cold flex temperature or low temperature impact resistance without recourse to high levels of plasticiser which would reduce hardness and tensile strength at higher temperatures to an unacceptable degree. Their key characteristic in PVC is a relatively flat curve of modulus of rigidity versus temperature. This is invariably associated with the plasticiser having a flat viscosity / temperature curve. These features are provided by linear flexible molecular structures found in esters of dibasic aliphatic acids typified by dioctyl adipate.

'Migration resistant'/'non-migratory'
This is a very broad classification since the relative performance of different plasticisers will depend on the nature of the liquid or solid into which they are migrating as discussed in Chapter 3. Whilst higher molecular weight polyester plasticisers do show a broad spectrum of migration resistance, complete resistance is shown only by plasticising polymers. These materials which have very much higher molecular weights than conventional plasticisers are discussed at the end of this chapter.

'Oil-resistant'/'solvent-resistant'
These are more specific descriptions which are often applied to polyester plasticisers. Their good performance here depends on their high molecular weight which prevents rapid diffusion through the PVC structure. In addition to this kinetic effect, the high polarity of the polyester chain units favours retention by PVC in competition with dissolution in non-polar extractants.

'Non-marring'
This term is used to describe plasticisers which are resistant to migration from plasticised PVC products into other polymers with which they are in contact. The main plasticisers falling into this category are trimellitates and some polyesters. Illustrative examples are given in Chapter 3, Section 3.6.

'Flame – retardant'
Phosphate esters and chlorinated paraffins are traditionally described as flame retardant plasticisers but this is a misnomer in the context of PVC. Whilst these plasticisers increase fire resistance when added to readily combustible polymers, unplasticised PVC is intrinsically resistant to burning as a result of its high chlorine content. Plasticisation with phosphates or chlorparaffins progressively **decreases** the fire resistance of PVC although to a much lesser extent than with other types of plasticiser (see Chapter 3, Figure 3.9).

'Stain resistant'

Butyl benzyl phthalate (BBP) is conventionally accepted in the flooring industry as being the standard stain resistant plasticiser for PVC. This is associated with its strong solvation of the polymer, a characteristic shared by di(propylene glycol) dibenzoate which is sometimes used as a less volatile alternative to BBP. The stain resistance of plastisol-based products is enhanced by mixing BBP with low viscosity components which allow the use of low plasticiser levels. Proprietary mixtures of this type are marketed into the flooring industry.

'Stabilising'

Stabilising plasticisers have molecular structures giving them a combination of plasticising and heat stabilising activity with PVC. Established materials in his category are epoxidised compounds typified by epoxy soya bean oil. Their epoxy groups give a useful co-stabilising effect when they are used in conjunction with the metal soap type of stabilisers. In practice their plasticising function tends to be regarded as secondary. Some products contribute particularly good low temperature flex whilst others have high resistance to extraction. All are secondary plasticisers with limited PVC compatibility.

'Antistatic'

Antistatic plasticisers function as nonionic antistatic agents by facilitating the migration of ionic species across the surface of a PVC product (see Chapter 2, 2.4.11). However they have much higher compatibility with PVC than simple antistats and can constitute a significant part of the plasticiser system. Their effect is related to the presence of polyether groups and/or hydroxy groups in their structures. They may be simple chemical compositions or proprietary blends.

'Non-toxic'

This bold term was used in former times to describe some types of plasticiser suitable for food contact applications. Nowadays it would generally be regarded as too simplistic to be helpful.

'Polymerisable'

The so called 'polymerisable plasticisers' are really more akin to process aids since their action can be described as plasticisation only at the process stage. In their monomeric state they are liquids of low viscosity having structural attributes giving them compatibility with PVC. In this state they contribute to lowering the melt viscosity or plastisol viscosity of the PVC composition. During processing to the final product polymerisation of the monomer occurs resulting in the formation of a crosslinked interpenetrating network not involving any reaction with the PVC. This gives the composition reduced flexibility but also the enhanced toughness required for specific end uses. The scale of consumption of polymerisable plasticisers remains small. Examples of this technology are:

(a) Diallyl phthalate

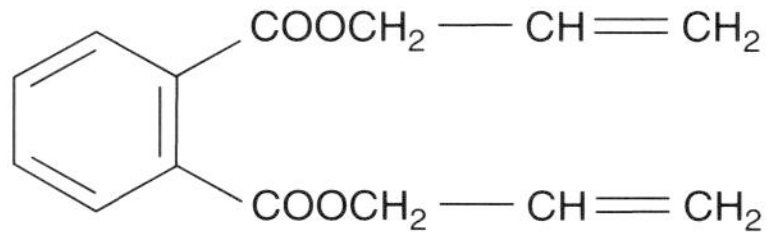

$$\text{COOCH}_2 - \text{CH} = \text{CH}_2$$
$$\text{COOCH}_2 - \text{CH} = \text{CH}_2$$

is used in plastisols for hard coatings with good adhesion to steel. Crosslinking requires a temperature of about 180°C and the use of a peroxide initiator. The type of PVC heat stabiliser present strongly influences crosslinking, mixed metal soaps being far less satisfactory than leads.

(b) Trimethylol propane trimethacrylate

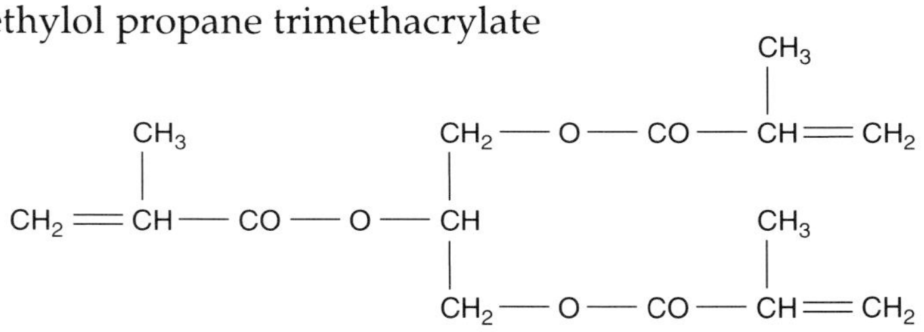

$$\text{CH}_2 = \text{CH} - \text{CO} - \text{O} - \text{CH} \begin{cases} \text{CH}_2 - \text{O} - \text{CO} - \overset{\displaystyle \text{CH}_3}{\text{C}} = \text{CH}_2 \\ \\ \text{CH}_2 - \text{O} - \text{CO} - \overset{\displaystyle \text{CH}_3}{\text{C}} = \text{CH}_2 \end{cases}$$

is used in thin walled automotive cable insulation to reduce hot deformation. Crosslinking is achieved by electron beam curing (usually off-line) following extrusion.

5.3   ESTERS AS PLASTICISERS – GENERAL

Nearly all plasticisers used in PVC or any other polymer are esters. In tonnage term the chlorparaffins are the only significant exception.

The reason for the pre-eminence of esters is the enormous variety of structures which can be produced from readily available raw materials using relatively simple and inexpensive process technology. In simple terms it is the ester groups which provide compatibility with PVC whilst the size and shape of the remainder of the molecule dictates plasticising efficiency and permanence.

The general process of esterification is the reaction between an acid and an alcohol with the elimination of water. In some cases it is necessary to use a more reactive or cheaper derivative of the acid, for example phosphorus oxychloride for phosphates, phthalic anhydride for phthalates. The following lists indicate the variety of raw materials which have been used for manufacture of plasticiser esters. They include acids with basicity up to 4 and alcohols with up to 6 hydroxy groups.

ACIDS

monobasic	*dibasic*	*tribasic*	*tetrabasic*
benzoic	orthophthalic	phosphoric	pyromellitic
capric/caprylic	(as anhydride)	(as oxychloride)	(as anhydride)
2-ethylhexanoic	isophthalic	trimellitic	
acetic	terephthalic	(as anhydride)	
butyric	adipic	citric	
alkane sulphonic	'nylonic'		
	succinic		
	sebacic		
	azeleic		

ALCOHOLS (AND PHENOLS)

monohydric
C1 to C13 (see Chapter 4 – Phthalates)
glycol ethers – e.g. methoxyethanol,
butoxyethanol, butoxyethoxyethanol
benzyl (as chloride)
phenols – phenol
    cresol
    xylenol
    isopropylated phenols

dihydric (glycols)
ethylene glycol
diethylene glycol
propan-1,2-diol
dipropylene glycol
butan-1,3,-diol
neopentyl glycol

trihydric	*tetrahydric*	*hexahydric*
glycerol	pentaerythritol	dipentaerythritol
trimethylolpropane		

5.4 CHLORINATED PARAFFINS

Manufacturer references (see Appendix 1): 10, 30, 32

Since the early nineteen sixties chlorparaffins used as plasticisers have been based on 100% straight chain paraffins produced by molecular sieve processes. This source provided a great advance on earlier feedstocks which gave products with poor thermal stability. Commercial chlorparaffins are produced from normal chlorparaffin fractions C10–C13, C12–C14, C14–C17 and C18–C20 and chlorine contents range from 30 to 70% by weight.[76] Of these it is the C14–C17 (average C15) fraction chlorinated to 45% and 52% which are the most important as PVC plasticisers.

In Europe PVC plasticisation is the main application of chlorparaffins accounting for more than half of their 85 kt/a consumption. They are also used as plasticisers for some types of paints, adhesives and sealants.

Chlorparaffins of high chlorine content are used as flame retardants (as distinct from flame retardant plasticisers) in polyolefins, unsaturated polyester resins and rubbers. Here the most effective grade is a solid product containing 70% chlorine. Following their use in PVC the largest outlet for chlorparaffins is as components of metal working lubricants and gear oils. Their function here is to maintain lubrication and prevent wear under conditions of extreme pressure. In contrast to the situation in Europe chlorparaffins are little used as PVC plasticisers in the USA.

These materials are produced by passing chlorine gas through the liquid paraffin at a temperatures of approximately 100°C. Production of each tonne of chlorparaffin generates approximately half a tonne of hydrogen chloride. Economical operation therefore depends on the availability of outlets for this byproduct, for example as a feedstock (via oxychlorination) for production of vinyl chloride monomer.

In order to prevent the release of hydrogen chloride resulting from slight decomposition in storage, chlorparaffins are usually supplied containing a small proportion of an epoxy compound as an HCl scavenger.

The world's largest supplier of chlorparaffins is ICI with plants in various European locations as well as Canada, Australia and Thailand. Very similar materials are available from a number of other producers including Caffaro, Italy. As organochlorine compounds the chlorparaffins have attracted particular attention because of the range of environmental concerns mentioned in Chapter 9, 9.8. This has had the effect of creating uncertainty in the future of the market which has influenced some producers to withdraw rather than invest in plant renewal.

Chlorparaffins are secondary plasticisers for PVC, that is to say they would normally have to be used in mixtures with other plasticisers because of their relatively low compatibility limits with the polymer.

There are two main factors accounting for the volume used in PVC, one technical, the other purely economic. As stated earlier, whilst the incorporation of any plasticiser into PVC detracts from its fire resistance this effect is much smaller for chlorparaffins than for nearly all other classes of plasticiser (see Chapter 3, Figure 3.7). Hence they present a widely used option for achieving a compromise between flexibility and fire resistance.

With the exception of certain hydrocarbons which have very limited use, chlorparaffins are the cheapest of all PVC plasticisers. Historically their price in many markets has tended to track that of general purpose phthalates, typically being 70% of DOP price. Consequently they are often used in PVC simply as extenders with the primary aim of reducing cost rather than contributing any technical benefit. When formulations are costed on the basis of volume, the cost advantages of using chlorparaffins are somewhat reduced by their relatively high density, which increases with increasing chlorine content. However generalisations of this type can be misleading and comparative costings always require case by case calculation.

	Density (kgl^{-1})
Chlorparaffins C14–C17	
40% Cl	1.10
45% Cl	1.16
52% Cl	1.25
58% Cl	1.36
(PVC 57% Cl	1.40)
Ester plasticisers	
DOP	0.98
DIDP	0.97
DOA	0.93

The viscosity of chlorparaffins increases roughly logarithmically with increasing chlorine content and increasing chain length. This places practical constraints on the composition of the variants which can be used with PVC. The effects of increasing chlorine content on various aspects of plasticiser performance are as follows:

solution temperature	decreases
plastisol viscosity	increases
compatibility limit	increases
softening efficiency	decreases
low temperature performance	decreases
volatility	decreases
general permanence	decreases
fire resistance	increases

As already stated most of the demand for chlorparaffin plasticisers for PVC is satisfied by two types:

(a) 45% chlorinate of C14–C17 n-paraffins.
Empirical formula C15 H27.2 Cl4.8

On average every third carbon atom in the chain is chlorinated.
Average molecular weight 377 (c.f. DOP 390)
Viscosity at 25°C 200 mPa.s (c.f. DOP 57)

(b) 52% chlorinate of C14–C17 n-paraffins
Empirical formula C15 H27.2 Cl6.3
Average molecular weight 420 (c.f. DINP 418)
Viscosity at 25°C 1600 mPa.s (c.f. DINP 75).

A further plasticiser based on the same feedstock and containing 58% chlorine is available.

Although the two major chlorparaffins closely have average molecular weights closely matching the general purpose phthalates they show higher volatile loss than DOP and DINP respectively. This is mainly because their components have a distribution of chlorine contents, a proportion having considerably lower molecular weight than the average. When used as sole plasticiser the compatibility limits with PVC are approximately 45 p.h.r. for the 45% chlorinate and 60 p.h.r. for the 52% chlorinate. These figures are typical for mechanically unstressed specimens containing no other liquid additives of low compatibility. Data of the type shown in Figures 3.1 (Chapter 3) and 5.1 are available from suppliers to aid formulation.[77] Figure 5.1 indicates the compatibility limits of chlorparaffins used in conjunction with various primary plasticisers.

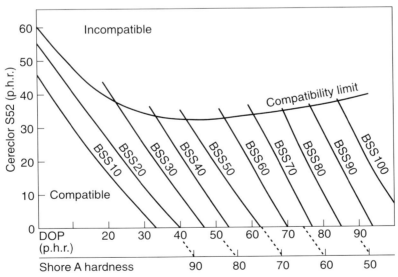

Fig. 5.1 Compatibility limits and softness contours for PVC plasticised with DOP and Cerector S52.

Figure 5.1 additionally shows contours defining compositions of equal softness plasticised with mixtures of DOP and the 52% chlorinate.

Applications of the two types overlap, although they have distinct areas of major use because of differences in viscosity and volatility. The 45% chlorinate can be used to produce plastisols having low viscosity and good viscosity stability (despite having higher viscosity per se than most primary plasticisers). Hence it is widely used in products made by spread coating such as wallpaper and coated fabrics. The 52% chlorinate which has much higher viscosity tends to be confined to melt processing. It has been widely used in cable compounds where the volatility of the 45% chlorinate would be too high to conform to heat loss specifications.

5.5 ESTERS OF LINEAR DIBASIC ACIDS

Manufacturer references (see Appendix 1): 3, 4, 5, 6, 13, 14, 15, 18, 21, 22, 23, 24, 27, 29, 31, 33, 35, 41, 45, 46

This familiar group of plasticisers consists of materials with the general structure:

$$ROCO(CH_2)_nCOOR$$

of which the adipates (n=4) are the most important. Other esters of this type which are known as plasticisers are:

succinates	n=2
glutarates	n=3
'nylonates'	n=2/3/4
azelates	n=7
sebacates	n=8

The flexible linear molecular structure of these materials gives them the common characteristics of low viscosity and good low temperature plasticising performance. Variation of molecular weight superimposes a wide range of volatility and PVC compatibility.

The same range of monohydric alcohols used as phthalate feedstocks are available for these esters. The use of longer chain linear alcohols is constrained by the tendency of the derived esters to crystallize at relatively high temperatures causing problems with handling and PVC incompatibility.

The reason for the leading position of the adipates within this group of plasticisers is the ready availability of adipic acid which is a commodity feedstock for nylon. It is a product of high purity manufactured principally by oxidation of cyclohexane.

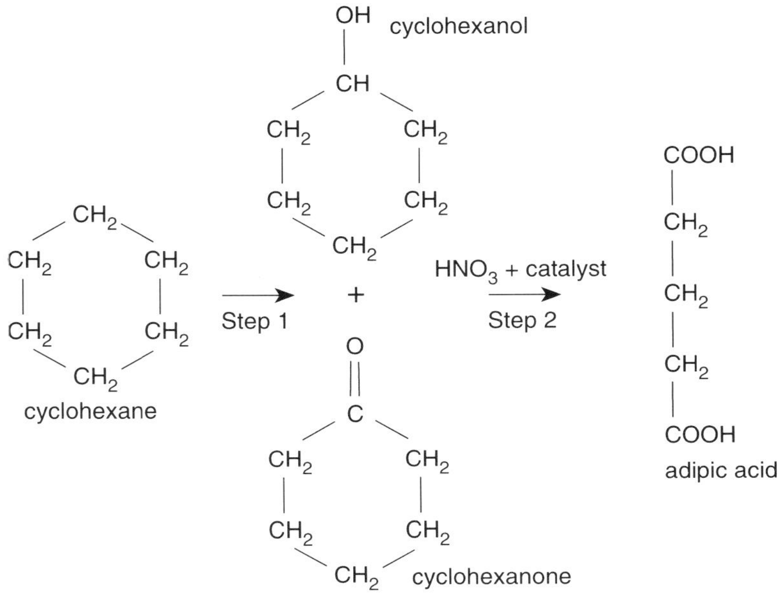

OH cyclohexanol

CH

CH₂   CH₂

CH₂   CH₂

CH₂

CH₂
CH₂   CH₂
CH₂   CH₂
CH₂
cyclohexane

Step 1   →   +   HNO₃ + catalyst → Step 2

O
‖
C
CH₂   CH₂
CH₂   CH₂
CH₂   cyclohexanone

COOH
CH₂
CH₂
CH₂
CH₂
COOH
adipic acid

In the second step over-oxidation gives a proportion of succinic and glutaric acids. Crystallization of pure adipic acid from the crude product leaves a mixture of adipic, glutaric and succinic acids which can be isolated as a useful esterification feedstock known as 'AGS acids' or 'nylonic acid'. A major outlet for this product is the dimethyl ester which is now well established as a slow evaporating solvent for surface coatings. Formerly C8 to C10 nylonates were marketed as PVC plasticisers usually having the status of lower priced alternatives to adipates. For a period in the early nineteen eighties they achieved significant sales as proprietary mixtures with commodity phthalates (Hexaplas® OPN, Hexaplas DPN) offering cost/performance advantages over linear phthalates for some applications.

'Pure' succinates and glutarates are of little commercial interest as plasticisers.

The longer chain dibasic acids, sebacic and azelaic are derived from natural sources. Of the two it is only sebacic acid which is of any significance as a plasticiser feedstock in Europe. It is produced from castor oil and relative to adipic acid its price is high and potentially more variable.

Comparative performance data for key adipates and sebacates are shown in Table 5.1. Notes on individual products of commercial and technical interest follow.

TABLE 5.1
PHYSICAL PROPERTIES AND PLASTICISING PERFORMANCE – DOA, DIDA AND DOS VERSUS DOP

	DOA	DIDA	DOS	DOP
Molecular weight	371	427	427	390
Viscosity at 25°C (mPa.s)	14	30	22	78
Solution temperature (°C)	149	164	160	124
PVC properties at 67 p.h.r.				
Shore A Hardness	70	75	75	73
Cold flex temperature – at 950 N/sq mm (°C)	−100	−95	−97	−68
– at 310 N/sq mm (°C)	−54	−56	−64	−36
Low temperature impact – no failure at (°C)	−63	−56	−64	−36
Loss in weight after 24 hrs at 90°C (%)	5.3	1.0	1.0	1.1

Source: Huls AG[78]

Dibutyl adipate and di-isobutyl adipate
Dibutyl adipate and di-isobutyl adipate are far too volatile for normal use with PVC and are restricted to specialised niche applications with other polymers.

Di-n-hexyl adipate
The main use of this material is as a plasticiser for polyvinyl butyral. Plasticised PVB is used as the interlayer in laminated safety glass. Di-n-hexyl adipate has higher compatibility with the polymer than adipates of higher molecular weight.

Di-2-ethylhexyl adipate
This is known conventionally as dioctyl adipate, DOA, although the abbreviation DEHA has often been used in recent years in the context of toxicological studies. DOA is used in the largest quantity of any of the linear diester plasticisers. It is a commodity product, being produced from commodity raw materials by a simple esterification process.

DOA can be regarded as the simple archetype low temperature plasticiser for PVC. In combination with general purpose phthalates it provides an economical way of meeting low temperature specifications. However its relatively high volatility excludes it from some applications. Although adipates in general are sometimes classified as secondary plasticisers, DOA has a fairly high compatibility limit with PVC. If lack of permanence is observed it is more likely to be due to its rate of migration than to incompatibility with the polymer.

DOA has long been used as a plasticiser in PVC cling film for food wrapping and this still constitutes its major application. It is now standard practice in Europe to use DOA in a mixed plasticiser system together with an approved polyester and epoxidised soya bean oil in order to contain the total migration within prescribed limits.

As a result consumption of DOA is at a lower level than formerly. During the nineteen eighties PVC cling film received a certain amount of public attention as a result of misunderstanding and misrepresentation of the toxicological risks associated with DOP. This subject is discussed elsewhere.

Linear C7–9 adipate
Two very similar variants of this type are available, one based on 80% linear C7/C8/C9 alcohols derived from SHOP olefins (see Chapter 4, 4.7.1), the other on C7/C9 alcohols with a slightly lower linear content. The second of these is better known in the USA and may appear in formulations of American origin. Relative to DOA these products have the advantage of lower volatility whilst avoiding the poor fusion and compatibility characteristics of the C9 and C10 adipates.

Adipates of C6/C8/C10 and C8/C10 Ziegler alcohols
Adipates of these 100% linear alcohol mixtures appear in the product lists of a number of European and American suppliers. They have the same advantages of linearity as discussed above for the C7–9 adipates.

Co-esters of the C8/C10 alcohol with branched alcohols are marketed and these avoid the problem high crystallizing temperature encountered with the 100% linear ester.

Di-isononyl adipate (DINA)
There are two different versions of this adipate on the market, these being produced from the same alcohols used for the two principal forms of C9 phthalate (see Chapter 4, 4.6.2 and 4.6.3). The alcohol based on polygas olefins contains components with an average of about two methyl groups per alkyl chain. The adipate of the alternative less branched alcohol has slightly better all round performance. Both products have the advantage over DOA of being significantly less volatile.

Di-isodecyl adipate (DIDA)
Di-isodecyl adipate has the highest molecular weight of any of the dialkyl adipate plasticisers in common use. Its high PVC solution temperature and low compatibility limit place it clearly in the category of secondary plasticisers. Its advantage over other adipates is its low volatility, comparable with that of DINP which makes it suitable for use in applications subject to moderately high temperatures.

DIDA has exactly the same empirical formula ($C_{26} H_{50} O_4$) as DOS (see below). As shown in Table 5.1 the two plasticisers have generally similar performance although DIDA is somewhat inferior on key points. DIDA has the ongoing advantage of considerably lower price.

Adipates of glycol ethers
The most important plasticiser in this small tonnage category is the adipate of butoxyethoxyethanol, $C_4H_9OCH_2CH_2OCH_2CH_2OH$ known as di(butylcarbitol) adipate. It can be processed with PVC with which it shows a moderate level of compatibility, in addition conferring useful antistatic properties. However its principal application is not in PVC but as a speciality plasticiser for nitrile rubbers where it shows a valuable combination of low temperature performance and low volatility.

Dibutyl sebacate (DBS)
This is one of the few plasticisers established in polyvinylidene chloride film (see Chapter 7, 5.7) where it is used at low addition levels to aid extrusion. DBS has also long been used in small quantities as low temperature plasticiser for some synthetic rubbers.

Di-2-ethylhexyl sebacate = dioctyl sebacate (DOS)
Dioctyl sebacate is well known in PVC technology as perhaps the best available low temperature plasticiser. Like DIDA which has the same empirical formula, DOS has low volatility for plasticiser of this type and can therefore be used in products operating over a wide temperature range. In low temperature performance and chemical stability it is superior to DIDA.

Because of its high price its use in PVC is confined to highly specified products such as low temperature cable insulation for military applications. DOS also appears as a low temperature plasticiser in long established nitrile rubber formulations.

Because of its combination of low viscosity, low pour point, low volatility and good thermo-oxidative and hydrolytic stability, DOS is conventionally used as a component of lubricant formulations.

5.6 EPOXY COMPOUNDS

Manufacturer references (see Appendix 1): 1, 4, 27, 29, 44

These are compounds with structures containing oxirane rings –

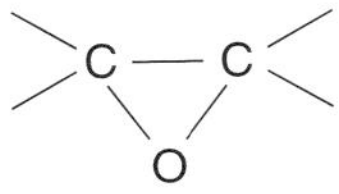

The types used as PVC additives are produced by epoxidation of unsaturated esters by reaction with peracids. The esters are either naturally occurring triglyceride mixtures such as epoxy soya bean oil, or monoesters produced from natural product derived monobasic acids.

These two groups are normally described as 'epoxy oils' and 'epoxy esters' respectively. They have the types of structure shown below.

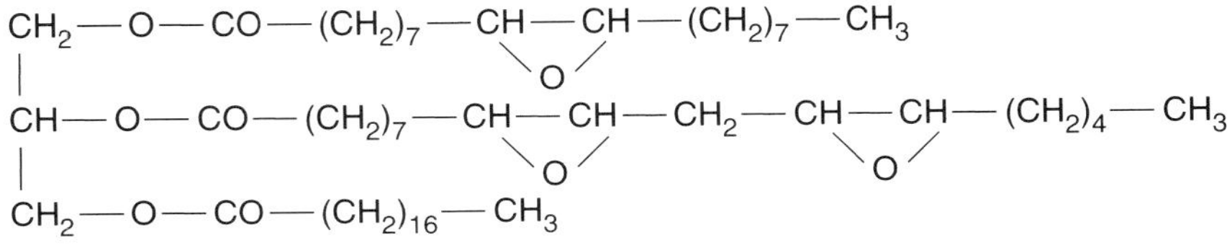

component of epoxidised soya bean oil

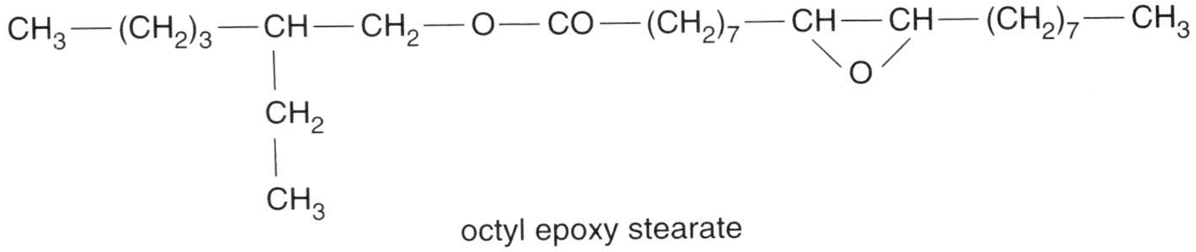

octyl epoxy stearate

These compounds have a dual function in PVC acting both as co-stabilisers (usually in conjunction with mixed metal systems) and as plasticisers. In the marketing of these products emphasis is usually placed on their stabilising function, their plasticising activity being secondary or even incidental. Heat stabilising efficiency is related to oxirane oxygen content and this parameter is included in sales specifications. The mod-ification of structure brought about by epoxidation also improves the PVC compatibility and permanence of the esters to an extent sufficient for them to be used as plasticisers.

The level of consumption of epoxy additives places them high in the league table of non phthalate plasticisers. However this simply reflects the fact that they are conventionally included as components of the stabiliser system for a very wide range of flexible PVC applications.

Typical levels of incorporation are around 6 p.h.r. at which they make a significant contribution to plasticisation and differences in performance between the different types become apparent.

Epoxy oils

The vegetable oils used as raw materials for these products are esters of glycerol with mixed C18 acids having a high level of unsaturation. The main such feedstocks are soya bean oil and linseed oil. The epoxidised products have molecular weights around 1000 which is well above those of phthalates and falls into the range of the lower molecular weight polyester plasticisers. They share with the polyesters the characteristics of high viscosity (500 mPa.s at 20°C for epoxy soya bean oil), low plasticising efficiency and high resistance to migration. Epoxy soya bean oil is used for example in PVC food packaging films where its low contribution to migration helps to achieve conformance to the prescribed maximum limit for total migration of all additives.

Epoxy esters

Epoxy esters are produced from esters of synthetic alcohols (typically 2-ethylhexanol) and C18 unsaturated acids. These acids are either pure oleic acid or the 'tall oil' acids which are derived from wood processing and the end products are termed 'epoxystearates' and 'epoxytallates' respectively. Their linear structures give them low temperature performance comparable with speciality low temperature plasticisers. Molecular weights are much lower than for the epoxy oils but still high enough (ca. 400 for octyl epoxy stearate) for their volatility to be no higher than the general purpose phthalates. Their low viscosity is of value in formulating plastisols.

The compatibility limits of the epoxy plasticisers classifies them as secondary plasticisers. In practice safe levels of incorporation are much lower than might be determined in simple tests. The predominant heat stabilising reaction of these additives is binding of hydrogen chloride by opening of the oxirane ring.

The epichlorohydrins produced in this reaction have low PVC compatibility and can cause surface exudation under some conditions. A further

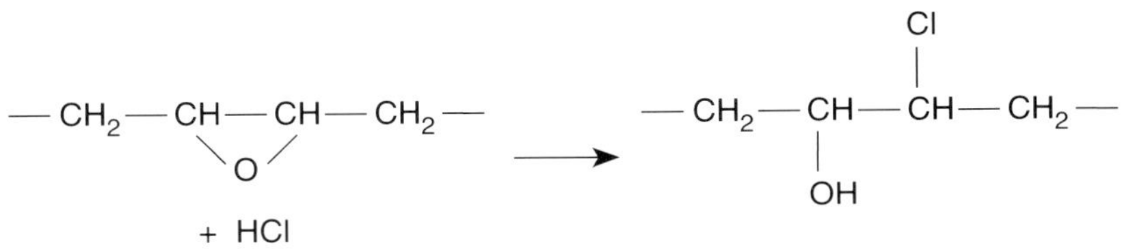

$$-CH_2-CH-CH-CH_2- \qquad \longrightarrow \qquad -CH_2-CH-\overset{\overset{\displaystyle Cl}{|}}{CH}-CH_2-$$

(with $-CH-CH-$ bridged by O; $+ HCl$ below left structure; OH below the second carbon at right)

+ HCl

undesirable potential reaction is polymerisation involving the epoxy groups to give an incompatible non-plasticising polymer.[79]

5.7 POLYESTERS

Manufacturer references (see Appendix 1): 1, 3, 5, 6, 7, 19, 22, 24, 27, 29, 34, 35, 42, 44, 46

Polyester plasticisers are produced by condensation of dicarboxylic acids with glycols. Hence their structures are built up of repeating units with regularly spaced ester groups. In the absence of other reactants the products consist of molecules with carboxyl and hydroxyl end groups and are described as non-terminated. A modification of structure which is often technically beneficial is achieved by including in the reactants a proportion of a monocarboxylic acid or, more commonly a monohydric alcohol as 'chain stoppers'. Thus the variables defining the composition of a polyester plasticiser are:

(a) the nature of the glycol, dicarboxylic acid and, if used the chain stopper.
(b) the degree of polymerisation, hence the weight average molecular weight.
(c) the molecular weight distribution.

Because they lack simple chemical identities these materials are marketed as proprietary products with only general descriptions of their chemical nature; e.g. 'polymeric plasticiser derived from adipic acid and butanediol'. Molecular weight is usually not declared but in relative terms is indicated by viscosity. Polyesters are classified descriptively as low-, medium- or high viscosity products. Molecular weights generally fall in the range 850 to 3,500 with corresponding viscosities ranging from about 500 to more than 10,000 mPa.s at 20°C.

An example of a typical polyester structure is shown below. This represents the average composition of a low viscosity 'modified' poly (propylene glycol adipate) chain stopped with 2-ethylhexanol and having a molecular weight of 1200.

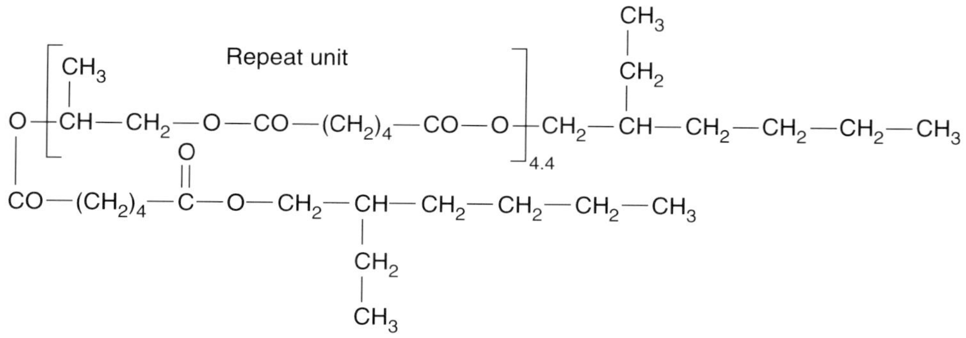

$$O-\left[CH(CH_3)-CH_2-O-CO-(CH_2)_4-CO-O\right]_{4.4}-CH_2-CH(CH_2CH_3)-CH_2-CH_2-CH_2-CH_3$$

It is desirable for the molecular weight distribution of these products to be as narrow as possible since lower fractions detract from migration resistance and higher fractions contribute to high viscosity. Control of this factor is achieved through appropriate process technology.

It is possible to produce polyester plasticisers with valuable performance characteristics by using a lactone as a raw material.[80] Lactones are cyclic esters in which the ester group is part of the ring structure. By ring opening under esterification conditions a lactone can be incorporated into a polyester chain. Each molecule incorporated takes the place of one glycol/dibasic acid repeat unit in the type of structure shown above.

The relationships between structure and performance of polyesters is more complex than for monomeric plasticisers. Since they are marketed as performance products and precise composition details are not readily available a detailed discussion of the subject here would be of little practical value. Once it has been decided that polyesters are appropriate options for a particular application then guidance on selection from suppliers becomes essential. As an indication of the variety of performance available from commercial polyesters, Table 5.2. shows the figures quoted by one manufacturer for a range of nine products meeting a wide range of market requirements. Comparative figures for a general purpose plasticiser, DIOP are included.[81]

Polyesters have about 2% of the total plasticiser market and are used in applications where specifications impose limits on levels of migration into solvents, oils and oily media. These include specialised types of cables and coated fabrics. A large part of the demand for polyesters is created by the need to avoid adulteration of foodstuffs by migrating plasticisers. Hence they are major constituents of formulations for food packaging films. The products which can be used here are strictly controlled by EU food contact regulation and are limited to: poly(propan-1,2-diol adipate) with or

TABLE 5.2
PERFORMANCE RANGE OF COMMERCIAL POLYESTER PLASTICISERS

	Polyesters	DIOP
Viscosity at 25°C mPa.s	7 to 32000	72
BS Softness number	24 to 34	45
Cold flex temperature °C	0 to −7.5	−20
Volatile loss %	2.9 to 12.0	22.4
Extraction by		
− hexane %	0.7 to 15.1	35.0
− mineral oil %	0.7 to 6.0	16.0
− water %	0.1 to 2.3	0
− 1% soap solution %	3.7 to 11.6	13.0

without an approved chain stopper and poly(butan-1,3-diol adipate) with or without an approved chain stopper

5.8 PHOSPHATES

Manufacturer references: 5, 24, 30, 31, 35

5.8.1 GENERAL

Phosphate plasticisers are esters of phosphoric acid and have the general structure:

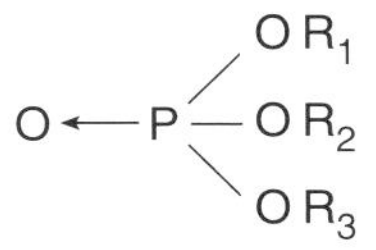

$$O \leftarrow P \begin{array}{l} OR_1 \\ OR_2 \\ OR_3 \end{array}$$

where R1, R2 and R3 are alkyl or aryl groups. Almost 90% of consumption is of triaryl phosphates which are produced from phosphorus oxychloride and various phenols. These are followed by alkyl aryl phosphates in which R1 and R2 = aryl and R3 phenyl. A third and much less significant group is the trialkyl phosphates.

Worldwide consumption of phosphate plasticisers is of the order of $70 \, \text{kta}^{-1}$.

Phosphates have a long history of use as plasticisers dating from the early part of the twentieth century when tricresyl phosphate was one of the first products to be substituted for camphor in nitrocellulose.

They were subsequently found to be effective PVC plasticisers and were adopted along with phthalates into the new PVC technology. Since they were incapable of matching the cost-effectiveness of phthalates for general purpose use they gradually acquired their current status as speciality plasticisers used mainly for their fire resistance.

Whilst their major application is in plasticising of polymers, phosphate esters are also used as fire resistant hydraulic fluids and as antiwear additives in lubricants. Here the phosphates of tertiary butyl phenol mixtures are particularly advantageous because of their extreme resistance to oxidation at high temperature.[82]

5.8.2 TRIARYL PHOSPHATES

The simplest member of this group, triphenyl phosphate is a crystalline solid with a melting point of 49°C and a low limit of compatibility in PVC. It is used in relatively low tonnage as a flame retardant plasticiser for cellulose acetate and some other polymers. The triaryl phosphates used in PVC are liquids produced from methyl phenols (coal tar – derived cresol or xylenol) or from isopropylphenols (obtained by alkylating phenol with propylene).

For many years the main phosphate plasticisers were tricresyl phosphate, TCP (otherwise known as tritolyl phosphate, TTP) and trixylyl phosphate, TXP. These remain in use at the present time but in much reduced volume. The feedstocks for these phosphates must be free of

component of mixed phosphate produced
from isopropyl phenol/phenol

ortho-substituted isomers in order to avoid an unacceptable level of toxicity in the esters. By the nineteen sixties feedstock availability had fallen as a result of the decline in coal tar distillation. One solution to this shortage was the dilution of cresol with phenol to give cresyl diphenyl phosphate. The other was the development of phosphates based on isopropylated phenol mixtures. These were designed as technical matches for the traditional products but had significant advantages. They had reduced odour and colour, lower toxicity, better light stability and greater general consistency than the old products. The isopropylphenyl phosphates are now by far the most important type of phosphate plasticiser.

In addition to their fire resistance (see chapter 3, 3.12) the main performance characteristics distinguishing triaryl phosphates from general purpose phthalate plasticisers are:

- low PVC solution temperature/rapid fusion
- very high PVC compatibility/ extender tolerance
- poor low temperature properties resulting from the lack of linear components in their structures
- high resistance to extraction by hydrocarbons

Comparative data for representative materials are given in Appendix 2.

Triaryl phosphates are generally used in mixtures with other plasticisers thereby diluting their deficiency in low temperature performance. A long established large application for these phosphates (used in conjunction with chlorinated paraffins) is in fire resistant PVC conveyor belting. The decline of deep mining of coal in Western Europe in recent years has had an impact on this market.

5.8.3 ALKYL ARYL PHOSPHATES

There are two materials in this class, 2-ethylhexyl diphenyl phosphate and isodecyl diphenyl phosphate, best known as Santicizer® 141 and Santicizer 148 respectively. The presence of an alkyl group in the structure gives a considerable modification in plasticising performance relative to the triaryl phosphates, specifically in greater softening efficiency, much better low temperature properties and better light stability. However these are gained at the expense of reduced fire resistance (see Chapter 3, Figure 3.9) and somewhat lower permanence as shown in Table 5.3.[74]

TABLE 5.3
COMPARISON OF ALKYL ARYL PHOSPHATES WITH TCP
(Plasticiser level 67 p.h.r.)

	Santicizer 141	Santicizer 148	TCP
Shore A Hardness	69	71	79
Cold flex temp. °C	−39	−35	−10
Activated carbon volatility (24 hours 87°C) % plasticiser loss	7.4	3.7	1.0
Kerosine extraction % plasticiser loss	7.3	3.9	1.0

Alkyl aryl phosphates are used mainly in application requiring a combination of fire resistance and cold flex, for example tarpaulins for building applications and coated fabric for contract seating. They are also included in polychloroprene and nitrile rubber formulations (e.g. for conveyor belting and foamed insulation for pipes).

5.8.4 TRIALKYL PHOSPHATES AND HALOGENATED PHOSPHATES

Examples of these materials are tributyl phosphate and trichloroethyl phosphate. They are used in limited quantities as flame retardant plasticisers for cellulose acetate. Halogenated phosphates have wider applications as non-plasticising flame retardants for polymers.

5.9 TRIMELLITATES AND PYROMELLITATES

Manufacturer references 3, 4, 15, 16, 19, 21, 22, 31, 33, 42

5.9.1 TRIMELLITATES – GENERAL

Trimellitates are produced by esterification of trimellitic anhydride (TMA) with monohydric alcohols of the same types used to produce phthalates (see Chapter 4, 4.6–4.7).

TMA   trimellitate

Plasticisers are the largest outlet for TMA, a speciality intermediate for which Amoco Chemicals has long been the world's only significant producer. During 1994 Alusuisse comissioned a European-scale plant for this material. TMA is produced by oxidation of pseudocumene which is obtained by distillation from refinery aromatic streams.

pseudocumene

The trimellitates are closely related structurally to phthalates and not surprisingly there is a degree of overlap in their areas of application.

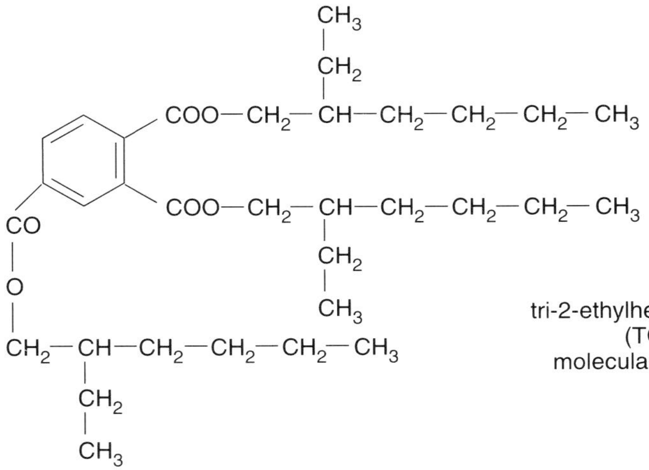

(TOTM)

tri-2-ethylhexyl trimellitate
(TOTM)
molecular weight 547

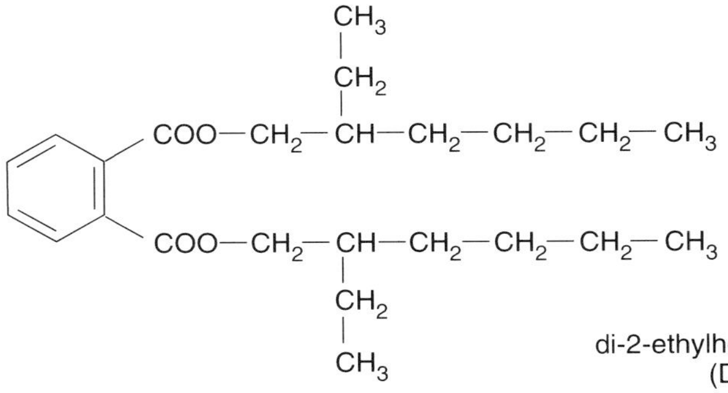

(DOP)

di-2-ethylhexyl phthalate
(DOP)
molecular weight 390

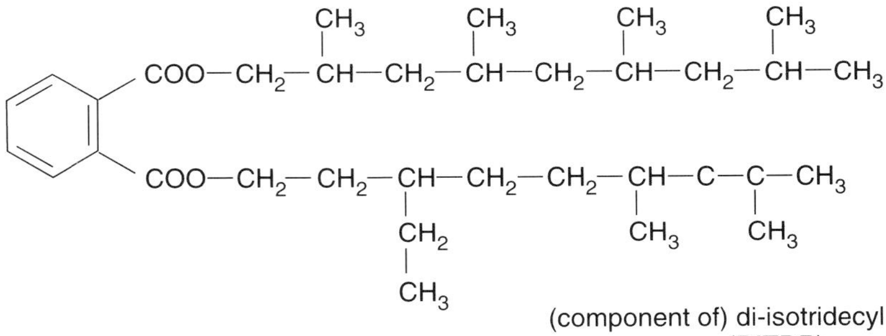

(DITDP)

(component of) di-isotridecyl phthalate
(DITDP)
molecular weight 530

TOTM, the most widely used trimellitate has a molecular weight comparable with DITDP which lies at the useful limits of PVC processability and compatibility. The presence of an additional ester group in the trimellitates avoids these limitations allowing materials with molecular weights up to about 600 to be used as plasticisers. Consequently trimellitates are able to take over where phthalates leave off in meeting specifications requiring low volatility.

When trimellitates first became commercially available there was a widespread assumption that their very low volatility would prove valuable in long life products used outdoors, particularly in hot climates. However this failed to take account of the reduction in PVC photostability caused by their relatively high UV absorption (see Chapter 3, 3.10) and the expected benefits were not realised.

Following their introduction in the nineteen seventies they became competitors not only with DITDP but also with polyesters in some applications. Whilst their molecular weights are lower than even the lowest polyesters they have the advantage of lacking the latter's molecular weight distribution. The lower molecular weight fractions of the polyesters can contribute disproportionately to volatile loss and in some cased to migration. The much lower viscosity of the trimellitates (TOTM 300 mPa.s at 20°C, c.f. 500–20,000 for polyesters) gives them considerable advantages in handling and processing. However trimellitates cannot match polyesters in their resistance to extraction by most solvents, oils and fats and this leaves major parts of the polyester market free of overlap.

The applications of trimellitates are associated mainly with their low volatility although their low level of migration into other polymers is also exploited in some cases. They have become established as the standard high temperature plasticisers for PVC, cable covering being their main market. For this purpose they are nearly always supplied containing an antioxidant, typically 0.5% of Bisphenol A. In the last few years they have also become widely used as plasticisers for car interior components which are exposed to high service temperatures, in particular crashpad (dashboard) skins. Here their resistance to migration into the foamed polyurethane backing material as well as their fogging performance exceeds that of phthalates.

Trimellitates in commercial use fall in a narrow range from C7 (average) to C9 (average). Below C7 they have little advantage in volatility over available phthalates and whilst individual products may have technically interesting combinations of attributes they have not yet prove cost effective versus alternative plasticisers. Above C9 limitations of poor PVC processability and low efficiency become apparent.

In addition to their main area of application in PVC trimellitates are used as high temperature plasticisers in some elastomers (e.g. chlorosulphonated polyethylene). Some trimellitates are useful as components of synthetic lubricants designed to operate over a wide temperature range.

5.9.2 Tri-2-Ethylhexyl Trimellitate, TOTM

As with the phthalates it is the 2-ethylhexyl ester which can be regarded as the standard of performance against which other trimellitates can be compared. TOTM provides a good all round balance of performance. It appears in the full range of end uses of trimellitate plasticisers including cables, automotive trim and miscellaneous applications requiring migration resistance. The fact that it is based on a commodity alcohol and is available from a number of suppliers usually gives it a competitive edge over most other trimellitates. Hence it is generally the preferred option unless another trimellitate has an over-riding performance advantage for a specific application.

The physiological effects of TOTM when used in PVC medical devices have been the subject of a significant body of research. It is now established in the USA and some European countries as a plasticiser for PVC blood bags and tubing. It is used in situations where lower levels of migration are required than can be used using DOP. For such applications a grade of high purity free of antioxidant is required.

5.9.3 Trimellitates of other Branched Alcohols

In contrast to the situation found with phthalates, the trimellitates of the polygas-derived oxo alcohols (iso-octanol and isononanol) have never gained a position in the European market. Tri-iso-octyl trimellitate has somewhat lower efficiency and markedly lower thermal stability than TOTM. Tri-isononyl trimellitate which is established in the USA, has very low volatility but its processing characteristics and plasticising efficiency are markedly worse than for TOTM.

5.9.4 Trimellitates of Linear Alcohols

n-C6/C8 trimellitate has advantages over TOTM in its easier processability (lower viscosity, lower PVC solution temperature) and higher plasticising efficiency. These occur at the expense of somewhat lower resistance to migration and lower electrical resistivity in compounds. A notable application of this plasticiser has been in PVC skins for automotive crashpads.

Tri-n-heptyl trimellitate has almost identical performance to n-C6/C8 trimellitate.

C7/C8/C9 trimellitate. This plasticiser is produced from 80% (+) linear Linevol® 79 alcohol derived from SHOP olefins (see Chapter 4, 4.7.1). It is less volatile than TOTM and confers better heat ageing resistance under extreme test conditions. In addition it has the same processing advantages as the two products described above although to a lesser extent. Trimellitates produced from **n-octanol** and **n-C6/C8/C10** alcohol are also available.

The most widely used linear alkyl trimellitate is produced from **C8/C10** Ziegler alcohols. This product has been familiar in the cable industry as a standard high temperature plasticiser for many years.

It has the best heat ageing resistance of any plasticiser in common use.

Its high temperature performance is combined with very good low temperature characteristics and it is therefore useful in products required to operate over a wide range of temperatures.

Table 5.4 compares the performance of a representative selection of trimellitates. Further comparative data are given in Appendix 2 and Chapter 3, 3.7.

TABLE 5.4
PERFORMANCE OF TRIMELLITATE PLASTICISERS

	TOTM	T79TM	T810TM	DOP
Molecular weight (average)	547	534	592	390
Viscosity at 20°C (mPa.s)	300	139	126	78
Gelation temperature(*) (°C)	128	125	135	109
B.S. Softness at 50 p.h.r.	19	24	13	34
Low temp. flex point at 50 p.h.r. (°C)	−15	−18	−28	−22
Heat ageing performance of compounds formulated to B.S. Softness 20: (14 days, 140°C, rapid air flow oven).				
Plasticiser level (p.h.r.)	50	47	58	−
Volatile loss (% of compound)	9.1	4.7	2.9	−
Retention of elongation (%)	80	106	98	−
Retention of tensile strength (%)	99	98	107	−

*The gelation temperature test gives results approximately 20°C lower than the more commonly quoted 'solution temperature' but gives the same indication of the relative performance of different plasticisers.

5.9.5 PYROMELLITATES

Pyromellitates are produced by esterification of pyromellitic dianhydride (PMDA), a high priced intermediate used in the production of polyimide resins.

When PMDA became commercially available in the nineteen eighties it was a natural candidate for evaluation as a plasticiser feedstock and small quantities of tetra-2-ethylhexyl pyromellitate (TOPM) were produced for test marketing exercises. Since it follows DOP and TOTM in a structural sequence it appeared to have potential advantages in migration resistance. However any performance advantages were insufficient to sustain interest

PMDA   pyromellitate ester

in pyromellitates in the face of their very high price and they have so far remained a technical curiosity.

It is likely that pyromellitates would be technically excluded from some application since their high UV absorption would be expected to result in poor photostability (see Chapter 3, 3.10)

5.10   ESTERS OF ISOPHTHALIC ACID AND TEREPHTHALIC ACID

Manufacturer references (see Appendix 1): 20

isophthalates   terephthalates

Di-2-ethylhexyl isophthalate (dioctyl isophthalate, DOIP) and di-2-ethyl-hexyl terephthalate (dioctyl terephthalate, DOTP) are known commercially as plasticisers in the USA but are little used, if at all, in Europe. They are isomers of DOP, compared with which they are claimed to have some advantages in specialised applications. Feedstock and production costs result in much higher prices than for DOP and other general purpose plasticisers. However an interesting development has been the proposal to use scrap terephthalate polyester (for example PET bottles) as a raw material to produce DOTP by the process of degradative transesterification.[83]

The term 'iso-phthalate' is sometimes used quite incorrectly as a shorthand generic description of the large tonnage branched chain alcohol phthalates derived from polygas olefins, i.e. di-isononyl phthalate, di-isodecyl phthalate etc. The fact that this occurs perhaps gives an indication of the relative obscurity of the esters of isophthalic acid.

5.11   ALIPHATIC ESTERS OF GLYCOLS

Manufacturer references (see Appendix 1): 18, 21, 31, 33, 45

Typically these plasticisers have linear structures resembling the aliphatic diesters described in 5.4 (adipates etc.) and similarly have low viscosity and good low temperature performance. However their PVC compatibility is lower and their price tends to be high. Consequently their use in PVC is very small.

Esters of triethylene glycol (for example the mixed caprate/caproate) are used as plasticisers for nitrile and polychloroprene rubber.

$$CH_2-CH_2-O-CO-(CH_2)_8-CH_3$$
$$|$$
$$O$$
$$|$$
$$CH_2$$
$$|$$
$$CH_2$$ (component of) triethylene glycol di-caprate/caproate
$$|$$
$$O$$
$$|$$
$$CH_2-CH_2-O-CO-(CH_2)_6-CH_3$$

At one time triethylene glycol di-2-ethylbutyrate was in large scale use as the major plasticiser for polyvinyl butyral sheet for safety glass interlayers. It has since been superseded by alternative esters (see Chapter 7, 7.3).

A rather different plasticiser which can be classified chemically as a glycol diester is 2,2,4-trimethylpentan-1,3,-diol di-isobutyrate, best known as Kodaflex® TXIB.

TXIB is produced by esterification of the monoester which is marketed as Texanol®. Texanol which is obtained by self condensation of iso-

$$CH_3-CH-CO-O-C-\overset{CH_3}{\underset{CH_3}{C}}-\overset{CH_3}{\underset{O}{CH}}-CH-CH_3$$
$$|\qquad\qquad\qquad|$$
$$CH_3\qquad\quad CH_3\; O$$
$$|$$
$$CO$$
$$|$$
$$CH-CH_3$$
$$|$$
$$CH_3$$

Molecular weight 286

butyraldehyde, is a well known coalescing solvent used in emulsion coatings.

TXIB shows unique performance characteristics in PVC plastisols as discussed in Chapter 2, 2.6.4. It is a useful component of plastisols formulated to give hard end products since it confers low viscosity at low levels of addition. It has been used widely in cushion vinyl flooring for this purpose, usually in conjunction with butyl benzyl phthalate. However the use of TXIB in flooring has declined greatly in the last few years because its high volatility causes a level of vapour emission from the end product which is no longer acceptable in some markets.

5.12   ESTERS OF POLYHYDRIC ALCOHOLS

Manufacturer references (see Appendix 1): 5, 17

By far the most important plasticiser in this category is **glycerol triacetate (triacetin)**.

$$CH_2-O-CO-CH_3$$
$$CH-O-CO-CH_3 \quad \text{Molecular weight 176}$$
$$CH_2-O-CO-CH_3$$

Triacetin is approved and used in many countries worldwide as the standard plasticiser in cellulose triacetate cigarette filter tow. Its function here is unusual since it does not involve the production of a homogeneous plasticised polymer composition. Triacetin is sprayed onto the fibres and causes their surfaces to soften. This results in the fibre-fibre bonding required for good filter performance.

Other esters of glycerol are known as plasticisers (e.g. the tributyrate) but their use is relatively insignificant.

Esters of pentaerythritol and dipentaerythritol have structures which confer valuable performance characteristics.

Hexa-esters produced from dipentaerythritol and mixtures of linear acids in the range C4 to C9 are excellent high temperature plasticisers for PVC.[45] They also have the merit of having relatively high resistance to oil extraction without the problematical high viscosity of the polyester plasticisers. However the high price of dipentaerythritol combined with a

$$\text{HO}-\text{CH}_2-\underset{\underset{\text{OH}}{|}}{\overset{\overset{\text{OH}}{|}}{\text{C}}}-\text{CH}_2-\text{OH}$$

pentaerythritol

$$\text{HO}-\text{CH}_2-\underset{\underset{\text{OH}}{|}}{\overset{\overset{\text{OH}}{|}}{\text{C}}}-\text{CH}_2-\text{O}-\text{CH}_2-\underset{\underset{\text{OH}}{|}}{\overset{\overset{\text{OH}}{|}}{\text{C}}}-\text{CH}_2-\text{OH}$$

dipentaerythritol

difficult esterification process has caused them to be eclipsed by trimellitates for all but a few small niche applications.

A material in this category with a most unusual combination of attributes is pentaerythritol tetrabutyrate (PTB). Performance data are compared below with corresponding figures for DOP and DOA.[84]

TABLE 5.5
PERFORMANCE OF PENTAERYTHRITOL TETRABUTYRATE

	PTB	DOP	DOA
Molecular weight	416	390	371
Gelation temperature (°C)	117	109	126
PVC compound properties, 50 p.h.r.:			
B.S. Softness	34	34	36
Cold flex temp (°C)	−7	−22	−52
Volatile loss (% compound)	0.9	0.8	1.9
Plastisol viscosity at 60 p.h.r.: (mPa.s)			
after 1 day	1.8	6.9	1.8
after 7 days	2.1	9.1	2.3

PTB combines very low plastisol viscosity with a moderately low gelation temperature. This is the same unusual combination found in TXIB and is presumably due to the same structural factors (i.e. steric shielding of the polar ester groups preventing solvation of the PVC at ambient temperature).

What is even more unusual is that this combination is found in a material with fairly low volatility. The potential value of such a material to plastisol processors is clear. However because of the costs of feedstocks, specialised esterification requirements and toxicological testing it remains simply an interesting technical model.

The foregoing is something of a technical digression from the focus of this chapter on plasticisers of current commercial importance. It illustrates the point made early in the chapter that variation of ester structure continues to allow scope for tailoring plasticiser performance to specific end uses.

5.13   ALKYL SULPHONATE ESTERS

Manufacturer reference (see Appendix 1): 5

There are only two very similar materials in this category. They are phenyl esters of sulphonated n-paraffins and are produced by Bayer AG as Mesamoll® and Mesamoll II.

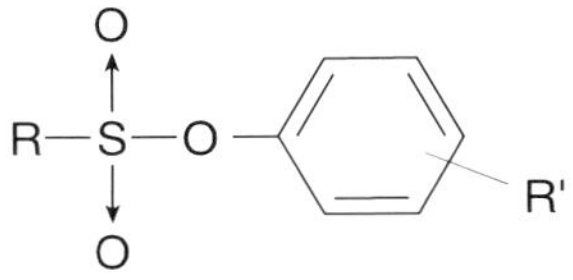

$$R-\underset{\displaystyle O}{\overset{\displaystyle O}{\underset{\displaystyle \|}{\overset{\displaystyle \uparrow}{S}}}}-O-\text{C}_6\text{H}_4-R'$$

R' = H and CH$_3$
R = C$_{15}$H$_{31}$ average in Mesamoll

Mesamoll has been used as a plasticiser since the nineteen thirties. Following the Second World War the same material was manufactured in the German Democratic Republic as Leuna ML. Mesamoll II is a variant introduced relatively recently. It has the same general characteristics as the Mesamoll but the use of a higher n-paraffin feedstock gives it the advantage of lower volatility. It is therefore more capable of meeting specifications requiring low volatile loss or fogging.

Were it not for their higher price these materials could easily be used as alternatives to the general purpose phthalate plasticisers in a wide range of applications. Their structure gives them some notable advantages over phthalates for certain processes and aggressive service environments. Their highly polar structure results in faster PVC fusion rates than can be achieved by other plasticisers of similar molecular weight (and hence similar volatility). In addition the phenyl sulphonate ester group has a low susceptibility to hydrolysis. This gives the Mesamolls high resistance to degradation during exposure to the weather, micro-organisms or alkaline media. Saponification rates for Mesamoll and DOP as determined by DIN 53404 are shown in Figure 5.2.[85]

In addition to its applications in PVC Mesamoll is used in a variety of other polymers where its processability, compatibility and chemical resistance are found to be of advantage. One large application is as a

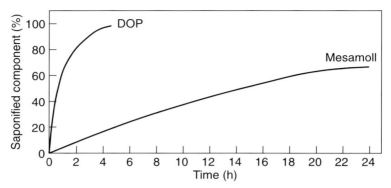

Fig. 5.2 Saponification rate of Mesamoll as compared to DOP (based on DIN 53404, Source: Bayer AG).

plasticiser in polyurethane direct glazing sealants for the automotive industry.

5.14   CITRATES

Manufacturer references (see Appendix 1): 18, 36

These are esters of citric acid, a raw material manufactured from sugars by enzymatic reactions.

$$
\begin{array}{l}
CH_2-COOH\\
|\\
HO-C-COOH\quad\quad \text{citric acid}\\
|\\
CH_2-COOH
\end{array}
$$

Citric acid is the only readily available aliphatic tricarboxylic acid. It can be used to manufacture esters with a wide range of molecular weight and polarity. The literature issued by Morflex, the main U.S.A. producer of citrates list approximately 20 such materials. In some of these the hydroxyl group of citric acid is acylated (to give the acetyl or butyryl derivative) and in other cases it remains unesterified.

Citrates are relatively expensive and whilst some of them show a useful balance of performance characteristics they do not display any out-standing technical advantages over phthalate plasticisers. Hence they have never become very significant in tonnage terms. Commercially the most important material in this category is acetyl tributyl citrate (ATBC) which is available from a number of suppliers. It is compatible with a variety of polymers including PVC, PVdC and PVAc. A major use is in extruded PVdC food packaging film.

$$CH_3-COO-\underset{\underset{CH_2-COO\,CH_2-CH_2-CH_2-CH_3}{|}}{\overset{\overset{CH_2-COO\,CH_2-CH_2-CH_2-CH_3}{|}}{C}}-COO\,CH_2-CH_2-CH_2-CH_3 \qquad \text{acetyl tributyl citrate} \\ \text{molecular weight 402}$$

Various citrates of higher molecular weight are offered as PVC plasticisers including –

acetyl tri-n-hexyl citrate
acetyl tri-2-ethylhexyl citrate
butyryl tri-n-hexyl citrate

The last of these has received particular attention following its evaluation as an alternative to DOP in PVC medical devices.[86]

The general image of citrates as plasticisers owes much to the perception that they are toxicologically benign. This is founded, at least in part on the common knowledge that citric acid is a natural product of low toxicity, occurring in citrus fruits and as human metabolite of carbohydrates. However, in relation to the large tonnage general purpose phthalate plasticisers which have been the subjects of huge programmes of toxicological testing, the various citrate esters used as plasticisers have been relatively little investigated. Any future comparison of toxicological risks is likely to require the generation of far more data than exist currently.

5.15   BENZOATES

Manufacturer references (see Appendix 1): 47

The consumption of benzoate plasticisers in Europe has so far been small although they are well known in the USA. There they have an established place in the PVC flooring industry where they are used as fast fusing stain resistant plasticisers. The most important benzoate plasticiser commercially is **di(propylene glycol) dibenzoate** which is broadly competitive with BBP but has the advantage of somewhat lower volatility.

Another established product with an interesting combination of technical properties is **2,2,4-trimethylpentan-1,3,-diol isobutyrate benzoate (Texanol® benzoate).**

⬡—CO—O—CH₃—CH—O—CH₂—CH—O—CO—⬡
│ CH₃ │ CH₃

Molecular weight 342
(c.f. BBP 312)

⬡—CO—O—CH₂—C—C—CH—CH₃ (with CH₃, CH₃, CH₃, O, CH—CH₃, CH₃ substituents)

Molecular weight 320
(c.f. BBP 312)

This plasticiser gives a combination of fast fusion, high stain resistance and relatively low plastisol viscosity.

Monobenzoates of higher aliphatic alcohols (e.g. C_{10}) are known in the USA as low viscosity components of plastisols (as alternatives to TXIB – see 5.11).

Perhaps more significant than their use in PVC is the fact that in the USA benzoates (in particular dipropylene glycol dibenzoate) have replaced much of the C_4 phthalate previously used in PVAc adhesives. This has not so far happened in Europe, partly because benzoates have been relatively expensive here and partly because to date there has been less H.S.E. pressure to avoid the use of phthalates. However the requirement imposed by the EC Dangerous Preparations Directive for labelling containers of 'preparations' containing dibutyl phthalate has now influenced this situation. More is said on this in Chapters 7 (7.2.2) and 9 (9.4).

At the time of writing it appears that benzoate plasticisers will shortly become available in the European market at far more competitive prices than hitherto. Aided by the 1994–1995 escalation of phthalates prices they could become as familiar in Europe as in the USA.

5.16   PLASTICISING POLYMERS

Manufacturer references (see Appendix 1): 5, 20, 26

Various flexible solid polymers are in use as components of semi-rigid or flexible PVC compositions. In order to achieve the same softness as with conventional plasticisation, it is necessary to use them in proportions comparable with that of the PVC resin. In some cases a proportion of a conventional liquid plasticiser is also included in the formulation. These polymers are generally much more expensive than PVC and they are usually confined to end uses where resistance to migration is paramount. The compounds are perhaps best regarded as specialised polymer blends rather than being classified as plasticised PVC.

Some types of plasticising polymer have very high molecular weight and require special attention to lubrication in melt processing. The following are examples of products in current use.

Trade name	Supplier	Composition
Elvaloy series	Dupont	Ethylene/vinyl acetate/carbon monoxide terpolymers. Mol. wt. ca. 230,000
Elvaloy HP series	Dupont	Ethylene/acrylate/carbon monoxide terpolymers. Mol. wt. ca. 400,000
Baymod L2418	Bayer	Ethylene/vinyl acetate copolymer (68% vinyl acetate)
Baymod PU	Bayer	Aliphatic polyester urethane
Chemigum P83	Goodyear	Partially crosslinked nitrile elastomer

CHAPTER 6

The Selection of Plasticisers for Specific Applications of PVC

6.1 INTRODUCTION

Selection of suitable plasticiser systems for different end uses is a task with varying degrees of difficulty. In some cases the technical requirements may be loosely defined and tolerant of a wide range of PVC compound compositions. At the other extreme the end product specification may place such conflicting requirements on the plasticiser system that complete conformance is impossible to achieve. In such cases it becomes clear that plasticised PVC is not a suitable material for the job in hand unless compromise on the part of the specifier is possible. For applications involving particular toxic risks – in food contact, medical products or children's toys – the formulator will usually be restricted to selection from a small group of approved plasticisers.

Regardless of the degree to which the target end properties are controlled by specifications, the formulator will always be seeking the lowest cost composition consistent with the needs for:

(a) processability   (Chapter 2)
(b) physical properties   (Chapter 2)
(c) permanence   (Chapter 3)
(d) conformance to any relevant regulations or codes of practice setting positive lists of approved additives and maximum permitted levels
                                                                              (Chapter 9)
(e) general health, safety and environmental considerations (Chapter 9)
(f) logistics of supply, storage and manufacturing.

The plasticisers covered in the previous two chapters have a wide range of prices, historically in the range from about 0.7 to more than $4\times$ that of PVC suspension polymer. Figure 6.1(a) shows approximate relative price per tonne of various types whilst Figure 6.1(b) makes the same comparison by volume. The latter is relevant for the processor buying raw materials by weight but making and selling products of a fixed volume. In both comparisons commodidity suspension PVC (as used for melt processing) is given unit value. Because of their relatively low density a number of plasticiser types are normally cheaper by volume than PVC and the cost penalty of using the more expensive specialities is reduced.

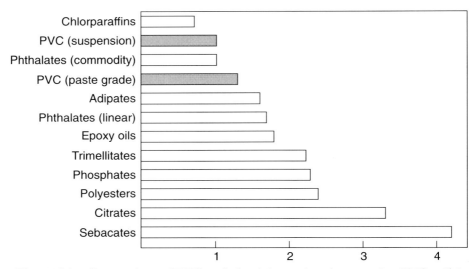

Fig. 6.1(a)   Comparison of PVC and plasticiser prices (suspension PVC = 1).*

If we take 55 p.h.r. as the average level of plasticiser found in flexible PVC products then typically the plasticiser becomes the main contributor to formulation cost when its price exceeds that of the PVC resin by a ratio of $100/55 = 1.82$. When plasticisers are cheaper than PVC then high addition levels give a cost benefit and this situation favours the use of plasticisers with low efficiency. For plasticisers more expensive than PVC high efficiency is desirable in order to contain costs.

Normally the first recourse of any formulator is to determine whether the required properties can be achieved using a general purpose plasticiser – DOP, DINP or DIDP. If one of these is already stocked in bulk then it will tend to be the preferred option. As already discussed in Chapter 4 these materials are broadly interchangeable in some applications although a requirement for low volatility leads to DIDP whilst an emphasis on processing speed favours DOP. For large tonnage operations, the cost savings obtainable by replacing a proportion of the phthalate by a chlorparaffin may be attractive.

* These comparisons are based on approximate average early nineteen nineties prices and are for illustrative purposes only. The unprecedented escalation of commodity phthalate prices during 1994 has caused a significant distortion of the historical pattern. However this is not expected to be a long term effect.

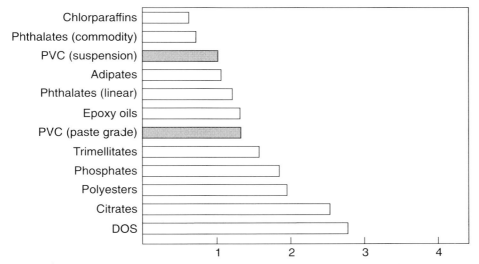

Fig. 6.1(b) Comparison of PVC and plasticiser volume costs (suspension PVC = 1).*

Should the phthalate alone be incapable of giving the required balance of properties then it becomes necessary to incorporate a proportion of speciality plasticiser(s) contributing the required characteristic(s). These have been discussed in detail in the last Chapter, examples being adipates for low temperature flex and phosphates for fire resistance. This approach is not always appropriate for addressing requirements involving plasticiser loss. As an extreme example, it would be a waste of money to partially replace DOP by a trimellitate as a means of preventing nitrocellulose marring since the DOP component of the blend would still migrate and cause the effect. However in some cases it is possible to arrive at the optimum cost/performance by starting from the other end of the performance spectrum – that is to select a speciality plasticiser which comfortably meets the end use specification requirements and then to dilute it to the maximum tolerable degree with a general purpose phthalate. This is a common approach to cost saving in high temperature cable insulation and is discussed further in the following section.

In opting for a multi-component plasticiser system the formulator has to take care that the benefit of one component is not cancelled out by the shortcomings of another. For example, suppose that from the starting point of a formulation containing 50 p.h.r. of DOP we need to reduce volatility without sacrificing fusion rate. A natural progression might lead through the steps:

(a) replacement of DOP with DIDP to reduce volatility
(b) partial substitution of DIDP by butyl benzyl phthalate (BBP) to compensate for the lower fusion rate of DIDP
(c) replacement of a further proportion of the DIDP by di-2-ethylhexyl adipate (DOA) to compensate for the poor cold flex given by BBP.

However the end result of these changes could be little or no improvement on the original formulation. Whilst DIDP is less volatile than DOP both DOA and BBP are significantly more volatile and although BBP has lower fusion temperature than DOP the converse applies to DIDP and DOA as shown in Table 6.1.

TABLE 6.1
Counterbalancing advantages(*) and disadvantages(**) of different plasticisers

	DOP	DIDP	BBP	DOA	L79P
Gelation temperature °C	109	123**	85*	126**	105
Compound properties, 50 p.h.r.					
B.S. Softness	34	23	40	36	37
Cold flex temperature °C	−22	−21	−5**	−52	−27
Volatility %	0.8	0.3*	1.6**	1.9**	0.4

An alternative approach to using a mixture here would be to replace DOP with a single plasticiser combining the required attributes such as a linear C_8 phthalate (L79P). Whichever approach is used there is likely to be a significant increase in cost relative to the starting formulation.

6.2 CABLES

Of the many applications of plasticised PVC it is the largest one, cable covering where the selection of plasticisers is structured to the greatest extent by specifications. These specifications may apply to the finished cables or cords or to the plasticised compounds used to extrude their insulation and sheathing. They may be issued by a variety of bodies:

(a) National Standards Authorities (or equivalent)
  e.g. British Standards Institution in the U.K
      Verband Deutscher Elektrotekniker (VDE) in Germany
      Underwriters Laboratories Inc. in the USA
(b) International bodies responsible for standards harmonisation
  e.g. la Commission International de Reglementation en vue
      de l'Approbation de l'Equipement Electrique (CENELEC)
      International Electrotechnical Commission (IEC)

(c)   Large individual purchasers of cables
        e.g. British Telecommunications PLC.

Conformance to the mechanical property requirements of these specifications does not generally place undue demands on general purpose
plasticisers except where extremes of temperature are encountered. Very
low temperature operation may necessitate the use of a low temperature
plasticiser such as dioctyl sebacate or di-isodecyl adipate. At high temperatures (as for example encountered in car engine compartments) the
physical limitation of plasticised PVC is thermoplastic deformation. There
are three ways of tackling this problem:

(a)   the use of a high *K*-value polymer (see Chapter 2, 2.3.2)
(b)   incorporation of a crosslinkable component (e.g. a 'polymerisable
        plasticiser' – see Chapter 5, 5.2)
(c)   plasticiser selection.

The plasticisers best suited to this situation are ones for which plasticising effect shows least dependence on temperature, in other words products which confer good low temperature flex at relatively low levels of
addition. This is illustrated by Table 6.2 which compares the hot deformation of compounds formulated to give the same cold flex temperature
using three different plasticisers. Hot deformation was measured according to B.S. 6469:198

TABLE 6.2

HOT DEFORMATION OF PVC COMPOUNDS WITH COLD FLEX TEMPERATURE $-35°C$

PLASTICISER	p.h.r	Hot deformation %
DOP	66	45
Tri-2-ethylhexyl trimellitate	76	52
Diundecyl phthalate	51	28

Cables are designed to be capable of operating at temperatures above
ambient for long periods without change of physical properties and this
has to take account of the potential for –

(a)   change of physical properties due to thermo-oxidative degradation
(b)   change of physical properties due to plasticiser loss.

Protection against (a) is ensured by the use of an adequate stabiliser
system including, if necessary an antioxidant to prevent initiation of
degradation through plasticiser oxidation. In order to counter (b) the
plasticiser volatility must be sufficiently low to prevent significant losses

during accelerated ageing tests. The plasticisers most commonly used in cables are arranged below in order of decreasing volatility.

Normal max operating temp.	*Branched phthalates*	*Linear phthalates*	*Trimellitates*	*Chlorparaffins*
70°C				
	DOP			C_{15},52% Cl
	DINP	C_8 average		
90°C				
	DIDP			
	DIUP			
105°C				
	DITDP	DUP		
			TOTM	
			T810TM	

KEY: DIUP = di-isoundecyl phthalate
DITDP = di-isotridecyl phthalate
DUP = diundecyl phthalate
TOTM = tri-2-ethylhexyl trimellitate
T810T = Linear C8–10 trimellitate

DOP is the most volatile plasticiser in large scale use in cable compounds but its performance is perfectly adequate for many large tonnage applications such as house wiring (at least in Europe) and appliance cords. Compared with DINP and DIDP it has the technical advantages of easier dryblend formation and greater acceptance of cheap chlorparaffin secondary plasticisers. In addition it is a more widely available commodity. However the material regarded as the standard general purpose plasticiser for the European cable industry is **DIDP**. This is because for a price which in recent history has been only slightly higher than DOP it is capable of satisfying a greater range of operating temperature requirements. Hence for compounders wishing to rationalise their bulk storage and compounding it becomes the obvious choice. DIDP has lower oxidative stability than DOP and in the absence of antioxidant this can lead to compound degradation at moderate test temperatures (ca. 100°C). Hence it is always supplied to the cable industry as a 'stabilised' grade normally containing 0.2–0.5% of Bisphenol A. All plasticisers of lower volatility than DIDP used in compounds tested at higher temperatures normally contain this or a similar antioxidant.

At the top end of the spectrum of heat ageing performance are the trimellitates. All but a few heat ageing specifications can be met using

TOTM and this has become widely accepted as the standard plasticiser used in high temperature PVC cable compounds. For the most extreme requirements, or if the formulator requires a greater margin of certainty in passing the specification, **T810TM** is available. For example this plasticiser is used in automotive cable required to retain good cold crack resistance after 240 hours at 150°C. For less extreme requirements it is not uncommon to reduce formulation costs by diluting trimellitates where possible with a cheaper phthalate. Where DIDP is in stock it is the obvious candidate for this purpose although greater levels of dilution can be achieved with higher molecular weight phthalates (see Chapter 4, 4.6.2).

DITDP has been used as a high temperature plasticiser in cables since before trimellitates appeared on the scene. There are instances where it still offers cost/performance advantages although it is a relatively difficult material to process. **DUP**, based on C11 alcohols with a high linear content, has a niche position in the cable industry as a result of its combination of heat ageing resistance and low temperature performance (see Table 6.2).

Most of the foregoing relates to avoidance of plasticiser loss through volatility. For specialised applications where cables are exposed to oils, fuels or solvents it may be necessary to use a polyester plasticiser or a plasticising polymer.

The hazards associated with PVC covered cables in fires have received considerable attention which has focused almost exclusively on their propensity for hydrogen chloride generation. A more balanced treatment of this subject places the various hazards of PVC and competing materials into proper context[87] in which for example the low heat release from PVC in fires is seen as a clear benefit. Formulation of fire retarded and low acid fire retarded grades of PVC compounds can invoke the various measures discussed elsewhere in this book. This may involve the use of phosphate and/or chlorparaffin plasticisers.

6.3 FLOOR COVERING

6.3.1 CUSHION VINYL FLOORING

Cushion vinyl (CV) flooring generally comprises four layers formed by spreading of plastisols in the following sequence:

Each of these component layers places its own requirements on the plastisol formulation either in processing or in end product performance; for instance :

encapsulation layer – low viscosity for complete impregnation and sealing of glass tissue

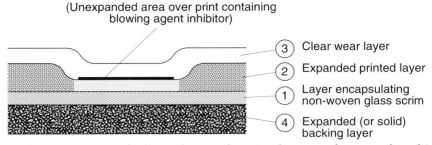

Fig. 6.2 Sequence of plastisol spreading in the manufacture of cushion vynyl flooring.

foamed layers – rapid fusion to ensure good cell structure
wear layer – clarity, toughness, stain resistance.

Ever since the advent of CV flooring such requirements have been satisfied by using three basic types of plasticiser in various combinations;

(a) a general purpose plasticiser, usually DOP
(b) a fast fusing/stain resistant plasticiser, nearly always butyl benzyl phthalate
(c) a low viscosity component, for example Kodaflex® TXIB. For a period in the nineteen eighties 'nylonates' filled this role (see Chapter 5, 5.5).

As discussed elsewhere (Chapter 2, 2.4.12 and Chapter 5, 5.11) the use of volatile low viscosity components in this application is no longer favoured. CV flooring producers have mainly dispensed with the use of such materials and aim to achieve low plastisol viscosity by selection of optimum PVC resins and fillers. This leaves **DOP** and **butyl benzyl phthalate (BBP)** as key raw materials but the position of both has been eroded to some extent by the promotion of **di-isoheptyl phthalate** in this market. Because this plasticiser is somewhat faster fusing than DOP this allows a reduction in the proportion of the more expensive BBP with consequent cost savings. Very recently **di(propylene glycol) dibenzoate** has appeared in the European market as a competitive alternative to BBP following the pattern established earlier in the USA.

6.3.2 PVC FLOOR TILES AND CALENDERED SHEET FLOORING

In hard 'vinyl' floor tiles the major component is calcium carbonate filler which accounts for up to 80% of the weight (ca. 65% by volume) of the product. Here the plasticised PVC functions as a binder. The need to achieve maximum dispersion of filler particles in the PVC imposes particular processing requirements on the formulation but these are readily achieved using a general purpose phthalate (most typically DOP) possibly with the incorporation of a chlorparaffin for additional cost saving. Likewise DOP is the plasticiser most commonly used in more flexible (less highly filled) tiles or continuous calendered sheet flooring. The higher plasticiser content of such products has in the past given problems with migration into flooring adhesives giving rise to shrinkage and adhesion. In some cases incomplete fusion has been found to be a contributory factor. It is likely to be more economical to minimise migration by adhesive selection than by the use of non-migratory plasticisers.

6.3.3 CONTRACT SAFETY FLOORING ETC

More specialised types of PVC flooring for industrial use, hospitals and public areas are covered by specifications which may contain clauses relating to fire resistance, antistatic properties etc. These are likely to require the use of speciality plasticisers or other additives which can add considerably to raw material costs. Some types of safety flooring are produced by spreading of plastisols containing an alkyl aryl phosphate as sole plasticiser. The demand for antistatic flooring has increased in recent years with the growth of the electronics industry. Antistatic plasticisers (Chapter 5, 5.2) add to the formulator's options for meeting specification for these products. At present consumption of such products remains small.

6.3.4 SOFT FLOORCOVERINGS

In addition to its use in smooth floorcovering, plasticised PVC is a major component in some types of carpets and carpet tiles, either as a pile adhesive or as a backing material. For such purposes it may solid or foamed and formulation usually contain the maximum tolerable level of filler. Application of the PVC to the substrate is carried out by plastisol spreading. Most commonly **DOP** as sole plasticiser is adequate for these plastisols although a fast fusing plasticiser (**butyl benzyl phthalate** or even a C_4 phthalate) is sometimes used to avoid the need for high process temperatures which might damage some substrates.

As with 'vinyl' flooring certain types of contract carpeting may require the use of specialty plasticisers.

Problems have been encountered when PVC-backed carpet tiles have been laid directly on new concrete. Hydrolysis of the plasticiser can give rise to a strong alcoholic odour (of 2-ethylhexanol in the case of DOP). The use of a hydrolysis resistant plasticiser would normally be uneconomic and the problem is best tackled by installing suitable barrier materials.

6.4 WALLCOVERINGS

Vinyl wallpaper with a wide range of decorative effects is produced by plastisol coating of paper. This industry grew rapidly in two distinct phases to become a major consumer of plasticisers. The first, in the late nineteen sixties saw the popularisation of flat wallcovering produced by reverse roll coating. At the same time the inhibition printing process well established in cushion vinyl flooring for embossing effects was applied to wallcoverings to give tile effects. In the second phase more than a decade later there was an almost explosive growth in the use of rotary screen printing for manufacture of textured foamed vinyl wallpaper. Many wallcovering manufacturers introducing screen printing opted to purchase ready formulated plastisols capable of meeting the particular rheological and foam expansion requirements of the process.

The end product requirements in large volume domestic wallcoverings do not impose any demands on plasticisers which cannot be satisfied by general purpose phthalates. Detailed selection of the optimum plasticiser can be strongly influenced by processing considerations. Both reverse roll coating and rotary screen printing subject the plastisol to extremely high shear rates. Specialised grades of PVC polymer are used in tailoring rheological properties to these conditions. Low viscosity plasticisers are seldom, if ever used in these applications. Reduction of viscosity by the use of volatile diluents (normally white spirit) is standard practice and this creates a need for measures to prevent airborne emissions.

The high surface-to-volume ratio of textured foamed vinyls gives potential for high plasticiser loss during processing adding to the problems of fume recovery and condensate disposal. This was one reason why the industry was generally receptive to the introduction of **DINP** as a less volatile alternative to DOP. When adjustments are made for the lower softening efficiency of DINP it gives lower plastisol viscosity than DOP allowing a reduction in the level of white spirit diluent. **DIDP** is normally found to be too slow in fusion for these applications particularly in foamed textured products where it gives inadequate expansion.

Whether DOP or DINP is selected it is often used in conjunction with a 45% chlorinated C_{15} paraffin (e.g. Cereclor® S45). This can improve viscosity and viscosity stability as well as reducing formulation cost. However if, as is now standard practice, a fume incineration system is installed as a means of preventing emissions, the hydrogen chloride produced from chlorparaffin combustion has potential for causing corrosion unless suitable materials of construction have been used.

Butyl benzyl phthalate is sometimes used as a fast fusing component in wallcoverings but far less commonly than in flooring.

6.5   CONSTRUCTION PRODUCTS AND TARPAULINS

A wide variety of flexible PVC products is used in weatherproofing and waterproofing applications. In extreme cases they are capable of retaining their performance after more than twenty years exposure to the weather. If the required service life is shorter or if the product is not to be exposed to strong sunlight and elevated temperature then the demands placed on the PVC composition will be less severe. As with any application where plasticiser permanence is a consideration, the surface to volume ratio of the PVC is an important factor, thin sections having the greatest potential for plasticiser loss. The great majority of single ply roof sheeting manufactured in Europe is manufactured from plasticised PVC. This type of product has an increasing share of the market for covering large flat roofs. The dominance of PVC in Europe contrasts with the situation in the USA where EPDM rubber is the main material for this application. Whilst PVC products have a long record of successful use and have some advantages in installation (ease of seam welding) an intrinsic weakness is the potential for plasticiser migration. Refurbishment of bitumen covered roofs requires the use of a barrier layer, typically in the form of a polyester fleece. PVC roof membranes are normally exposed to the weather but in some installations are covered by stone chippings. Under these conditions there have been instances where the rate of plasticiser loss has been unexpectedly high. Whilst the reasons for this are not always easy to diagnose fully, the problem has been attributed to microbiological degradation of plasticisers[88] and the use of biological stabilisers to counter this is normal practice.

The main plasticisers used in the upper (exposed) layer of PVC roof membranes are linear phthalates with average chain length C_{10}. They have long been accepted as the standard type for this application as a result of their thermo-oxidative stability, photostability low volatility and good low temperature performance. Linear phthalates are used for the same reasons in lighter weight PVC-coated polyester fabrics for large tent structures, sun shades etc. although the lightly branched type of C_9 phthalate has a major position here as a result of attractive cost/performance. Applications for this type of coated fabric extend into the transport sector where they appear in side covers on heavy goods vehicle trailers as well as traditional tarpaulin applications. Here cost pressures result in the use of commodity phthalates wherever possible. Where there is a need to combine weathering resistance with fire resistance an alkylaryl phosphate can be used in conjunction with the phthalate.

PVC products formed by calendering and extrusion are used for reservoir and tunnel lining and as waterstop profiles. The plasticiser used is nearly always a commodity phthalate, selection being dictated more by cost than by technical factors.

A completely different type of construction product consuming large quantities of plastisol is coil coated steel. This is used in the form of corrugated panels for cladding industrial buildings. The plastisol is applied by reverse roll coating as the protective top coat over the primed steel. PVC plastisols are regarded by this industry as a type of paint providing a flexible finish which is particularly resistant to mechanical damage during forming and installation. Typical coating thickness is about 0.2 mm which is unusually thick for a paint coating but thin relative to flexible PVC sheet products. Whilst the performance requirements for these coatings have some commonality with roof membranes there are significant differences. The purpose of the coatings is to retain the appearance of the panels and to protect the steel from corrosion during the guaranteed lifetime of the installation. The service performance of coated steel is known to be dependent on the whole coating system of which the PVC top coat constitutes only one part.

The high speed reverse roll application of thin coatings imposes enormous shear rates on the plastisols and their rheology needs to be tailored accordingly. Since a relatively hard finish is required plasticiser contents are typically no higher than 35 p.h.r. and volatile diluents are used to reduce viscosity. Selection of stabilisers and pigments are crucial for weathering resistance. For products used in hot climates plasticiser volatility becomes an important selection factor. Hence both processing and end use requirements have a bearing on plasticiser selection. In practice a variety of plasticiser systems are used in coil coatings including DIDP (stabilised), linear phthalates with average chain length C_8, C_9 or C_{10} and phthalate/adipate mixtures.

6.6 AUTOMOTIVE EXTERIOR APPLICATIONS

The largest of all markets for plastisol producers is underbody coatings and sealants for the automotive industry. This inconspicuous application constitutes by far the most significant application of PVC in car construction. Very few cars produced in Europe at this time do not incorporate these materials and the associated plasticiser requirement is of the order of 50,000 t.p.a.

Underseal is applied by airless spraying during the body painting sequence and is followed by spraying and curing of the finishing coat. Fusion of the plastisol is therefore carried out simultaneously with paint stoving. Hence underseal formulations have to be tailored to achieve full fusion on specific car painting lines. The performance requirements affecting the choice of plasticiser to varying degrees are thus:

(a) low viscosity at high shear rate together with a definite yield point (Chapter 2, 2.6.1) to prevent sagging between spraying and curing.
(b) good storage stability
(c) achievement of full fusion under specific oven conditions
(d) low volatility during fusion to prevent marring of the painted body by condensate drips.

Requirements (a) and (c) are usually achieved by measures other than plasticiser selection and depend mainly on the types of PVC resin used. Whilst some linear phthalates offer the optimum combination of attributes it is rare for underseal manufacturers to use plasticisers more expensive than commodity phthalates. DIDP and DINP are used in preference to DOP since they better fulfil conditions (b) and (c).

Following underseal, the next largest use of PVC on car exteriors is in extruded side strips which combine decorative and protective functions. Consumption of PVC in this application is not consistent and has tended to follow styling cycles. The extrusion compounds sold for the manufacture of these components normally contain a phthalate as sole plasticiser. Various phthalates, branched, lightly branched or linear have been used successfully in products meeting individual car manufacturers' specifications.

6.7   AUTOMOTIVE INTERIOR APPLICATIONS

Flexible PVC has long been used in a variety of components of car passenger compartments on account of its low cost, easy processing, durability and the range of design options which it offers. It was once the major covering material for car seats but was superseded by more comfortable textile materials. However it remains in use on a large scale for many other applications including:

– seat backs and sides
– crashpad (dashboard) skins
– sun visors
– door panels
– arm rests
– driver's foot well carpet panels
– water stop (inside the door casing)
– door frame beading
– gear stick gaiters.

A range of processes involving the forming of melts, plastisols and powders is used in the manufacture of such components. These include:

- calendering
- plastisol coating of fabrics
- plastisol coating of a textured release surface to produce unsupported expanded vinyl (UEV) sheet
- powder slush moulding

The last of these appeared in Europe in the mid nineteen eighties following its successful exploitation in Japan and the USA. By the use of powder slush moulding large single component crashpad skins having complex shapes and relatively soft leather grained surfaces are produced from plasticised PVC dryblends. The process allows designs which had not been achievable with older vacuum forming technology and these have been incorporated into an increasing number of new models. The formulation of powder compounds for these applications presented a major challenge in terms of both the moulding process and the performance of the end product in service. Nearly all of the slush moulding powder used in Europe is purchased from large PVC compounders rather than being produced in-house by the moulders. The majority of this compound has been produced under license from Japanese or American companies with longer experience of the technology.

There are parallels between the selection of plasticisers for car trim and their use in electrical cables. Whilst the type of plasticiser has some influence in all of the processes employed, processing considerations are not the main factor governing selection. This role is played by the component specifications issued by individual car manufacturers. As with cables it is volatility (in this case through its effect on windscreen fogging) which is the criterion giving the most important ranking of plasticiser performance.

The phenomenon of windscreen fogging and its measurement have been discussed in some detail in Chapter 3, 3.3.4. In Europe the most widely used fogging tests are those described in DIN 75 201. Car manufacturers set their own limits for maximum fog values for specific components as measured by these methods. The severity of each specification is related to the potential of the component to cause fogging. The most extreme conditions are encountered by the crashpad which is exposed to (filtered) solar radiation directly under the windscreen.

Of the plasticisers conventionally used in PVC trim DOP and tri-2-ethylhexyl trimellitate represent the extremes of fogging performance (see Chapter 3, Table 3.2). Materials plasticised with DOP would normally be expected to fail most fogging specifications based on the DIN test. However if they form part of a component which is effectively covered by a non-fogging material they may be acceptable. At the other extreme the relatively high price of trimellitates confines them to areas where temperatures are highest, that is to crashpad skins. One similarity to the cable industry is the use of DIDP as the major plasticiser for trim as a result of satisfying the largest number of specifications for least cost.

Higher linear phthalates, in particular the C_{9-11} products are commonly used in applications where the fogging resistance of DIDP is inadequate. They confer the additional advantage of superior cold crack resistance.

Conformance to any fogging specification requires the use of a plasticiser of appropriately high molecular weight. In addition it is required to be essentially free of volatile impurities. Control of this factor is discussed in the context of plasticiser quality in Chapter 8.

The high temperatures encountered below the windscreen provide not only the potential for release of fogging vapour but can cause physical deterioration of the PVC. Observed effects have included discolouration, exudation of incompatible material and embrittlement. The latter is a far too extreme effect to be explained by volatile loss of the plasticiser. In the construction of crashpads the PVC skins are back coated with a sprayed-on polyurethane composition which cures to give a semi-flexible impact absorbing foam. Plasticiser migration into this material during service is one cause of loss of flexibility of the PVC. Trimellitates have relatively high resistance to this process (see Chapter 3, Table 3.4).

The most serious deterioration of polyurethane-backed PVC has been found to be caused not by plasticiser migration from the PVC but by migration of amine catalysts from the polyurethane into the PVC.[89] These tertiary amines are potent dehydrochlorination catalysts and can initiate reactions leading to darkening and embrittlement of the PVC. The crosslinking which accompanies PVC dehydrochlorination is thought to reduce its compatibility with plasticisers thereby increasing their tendency to migrate. PVC degradation colour accounts at least in part for the phenomenon known as polyurethane 'staining' of PVC. The problems associated with polyurethane catalysts are now well recognised and have been countered by the use of reactive non-migratory catalysts in the PU and special stabiliser systems in the PVC.

Because of their low fogging and high resistance to migration the use of trimellitates (particularly tri-2-ethylhexyl trimellitate) in crashpads has become quite standard. One potential defect of trimellitates here is their known tendency to promote photodegradation which limits their useful-ness for outdoor applications. The fact that this does not appear to be a problem in practice may be due to the almost universal use of laminated glass for car windscreens. In this the PVB interlayer acts as a very effective screen against the damaging shorter wavelength solar irradiation.

6.8 PACKAGING

Although 890,000 tonnes of PVC was used in packaging in Europe in 1991 the great majority of this was rigid PVC for bottles and thermoformed foil.[90] The consumption of plasticised PVC in packaging is on a much smaller scale than is sometimes supposed. Its main use here is in extruded film for machine wrapping of fresh food and for sale as clingfilm. The other application is in closure seals (jar and bottle caps).

Historically the main plasticiser for PVC food packaging film has been di-2-ethylhexyl adipate (DEHA = DOA) used in conjunction with a proportion of epoxy soya bean oil which has the dual function of plasticiser and co-stabiliser. Film containing DOA proved to be ideal for this application because of its flexibility and stretch properties at a wide range of temperatures and its permeability to water vapour and oxygen. In addition the toxicology of DEHA was known in considerable detail. There are some variations between different European Union countries relating to the approval of plasticisers for this application. Such differences should eventually disappear when a positive list of approved polymer additives is incorporated into the laws of member states. In addition to being confined to this positive list, film producers will have to demonstrate that the 'global' migration of the additives present falls within a prescribed limit of $10\,mg/dm^2$. Specified extractants simulating different types of foodstuffs are used for testing. It is already known that the migration requirement can be met by the use of mixtures of DOA with certain polyester plasticisers (see Chapter 5, 5.6). Such mixtures have been in use for some time as a means of limiting the levels of DEHA migrating into food. This measure was initiated in response to rather confused concern over the safe use of DEHA during the nineteen eighties. This topic is discussed in its proper context in Chapter 9.

PVC food closure seals are produced from plastisols containing DOP, DINP or DIDP (without antioxidant). These seals may be either solid or cellular. The use of PVC in this application is now declining at the expense of polyethylene.

6.9 MEDICAL APPLICATIONS

Although the production of medical devices accounts for only 1% of PVC consumption in Western Europe, PVC is the material holding the largest share of this market (27% followed by polystyrene 18% and LDPE 17%).[91] Over 95% of medical grade plasticised PVC is used for the manufacture of containers (for blood, blood products and various aqueous solutions), flexible tubing and gloves, all of which are disposable items.[92] Factors favouring the use of PVC are its ease of processing, the range of physical properties available (by adjustment of plasticiser level) and its bio-compatibility together with low cost. To a greater degree than with consumer products any changes in raw materials for medical products are likely to be made solely on the basis of an unemotional and rational assessment of overall risk, benefit and cost.

As would be expected the range of plasticisers approved for PVC medical devices is extremely restricted. The plasticiser used in by far the greatest volume is **DOP**, usually referred to in any literature on the subject as **DEHP**. The European Pharmacopoeia has long included an entry for PVC plasticised with DEHP as a material for transfusion of blood and blood components. It remains the only plasticiser distinguished by such a reference. Because of its high compatibility with PVC and very low solubility in water, virtually no DEHP is extracted by medical aqueous solutions. However when lipid containing fluids are stored in or trans-ferred through flexible PVC plasticiser extraction can occur to varying degrees. The minute concentrations of DEHP extracted by the lipidic plasma fraction of blood have been found to be of positive benefit in stabilising the red cell membrane, thereby extending considerably the storage life of material stored in DEHP-plasticised bags.[93]

For some applications where there is a greater potential for extraction it may be considered desirable to use alternative plasticisers with better extraction resistance. Two such materials are currently of interest. One is tri-2-ethylhexyl trimellitate **TEHTM (TOTM)** which is used in haemodi-alysis tubing and blood platelet storage containers. The other plasticiser which has received attention is butyryl trihexyl citrate (**BTHC**) which has been shown to be a suitable alternative to both DEHP and TEHTM. Blood bags containing BTHC have been granted licenses in some countries. The use of these alternative plasticisers, in particular the citrate adds con-siderably to raw material costs in relation to 'standard' DEHP-containing products.

PVC surgical gloves are plasticised with either **DEHP** (DOP) or in some cases with **DINP**.

6.10   TOYS AND FOOTBALLS

PVC footballs, dolls and various soft squeezy toys are manufactured by rotational moulding of PVC plastisols formulated to give the required degree of hardness, flexibility or resilience for the application in question. In most cases DOP or DINP would be used as sole plasticiser although for hard dolls a low viscosity plasticiser system may be required in order to achieve adequate flow properties in combination with a low level of plasticisation. Hence these 'rigisols' are likely to contain plasticisers such as DIDA and Kodaflex® TXIB.

When PVC toys are used by very young children concerns arise over the risks associated with chewing the products. As regards plasticisers these are twofold. Firstly there is the risk of toxic effect through plasticiser extraction. Secondly solid fragments of the product could remain in the stomach after being bitten off and swallowed. Plasticiser extraction by gastric juices could then result in hardening of the fragments to a degree where they could cause physical damage. In Switzerland there is a strict and specific ban on the use of DOP in toys intended for use by the under-threes. This dates from 1986 when as a result of events discussed in Chapter 9 some legislators were singling out DOP for precautionary restrictions in the belief that it was a potential human carcinogen. Maintenance of the ban has not taken account of subsequent findings. The second risk is recognised in German Federal Health Office Recommendation XLVII which requires toys to be designed to eliminate the possibility of protruding portions being bitten off. However there is evidence that the hardening of plasticised PVC in the stomach is unlikely to occur.[94]

6.11 FOOTWEAR

Worldwide, plasticised PVC is one of the three major synthetic materials used for shoe soling (25% versus 28% for rubber and 24% for thermoplastic elastomers). In the UK its market share is higher at 35%. In addition to its major application in soling it is used for manufacture of some uppers, notably for wellington boots. Footwear production is the only large example of injection moulding of flexible PVC.

The advantages of PVC over competing materials for footwear are its low cost, ease of processing and the range of softness and flexibility available. The requirements for high flexibility in thick sections results in plasticiser levels being higher than for most other applications of PVC. On average the level is around 80 p.h.r. although it can be as high as 100 p.h.r. for wellington boot uppers. At these high levels plasticiser compatibility becomes an important consideration. Any surface exudation is particularly undesirable with unit soles which have to be bonded to the uppers with an adhesive.

With regard to plasticiser selection footwear compounds are seen as a technically undemanding application and it is rare for anything other than a commodity phthalate (**DOP** or **DINP**) to be used. Some chlorparaffin may be used as an extender although the scope for this is limited by compatibility considerations. Linear phthalates give some advantage in flex crack growth compared with the branched phthalates but their higher price is normally unacceptable in this market. Cost pressures and the technical tolerance of the application have resulted in footwear being regarded as a natural market for reclaimed or low specification phthalates.

In contrast to the situation described above the requirements for soling of industrial safety boots are technically demanding and result in higher raw material costs. Where PVC is used it is usually in an alloy with a plasticising polymer (see Chapter 5, 5.16) in order to meet specifications for oil resistance. There is some small specialised demand for antistatic footwear for use in environments presenting an explosion risk. Here the use of an antistatic plasticiser may be one of the options considered by the formulator.

6.12   CONVEYOR BELTING

The main use of heavyweight PVC covered conveyor belting is in underground coal mining. Since it was introduced in the UK in the nineteen fifties it has had an excellent safety record. One performance aspect distinguishing it from rubber (polychloroprene) belting is its failsafe behaviour in melting and fracturing under frictional heating of a stalled belt by a rotating drum thereby reducing the fire risk. Across the world the use of this type of belting remains significant although U.K. consumption has fallen greatly with the decline of deep mining.

The main factor governing plasticiser selection is conformance to extremely severe specifications for fire resistance. This is particularly difficult at the high plasticiser levels used to obtain the required flexibility in these thick section products. The plasticiser systems used here are predominantly mixtures of triaryl phosphates and chlorparaffins with a C_7 or C_8 phthalate as the third component.

In addition mineral fire retardants are included. A secondary influence on the choice of plasticiser is the need to tailor plastisol rheology and fusion to the specialised textile impregnation processes employed.

For other types of heavy duty belting used for surface conveyance of minerals fire resistance specifications are less restrictive and plasticiser systems contain much higher proportions of phthalates. Approved types of polyester plasticiser are used in lightweight conveyor belting for the food industry.

6.13 PROTECTIVE GLOVES

This is a relatively minor market for PVC involving a small number of European manufacturers. Whilst PVC has a cost/performance advantage over competing materials in this application, an increasing share of the market is being taken by gloves produced by dip coating nitrile rubber latex. These escape concerns over the environmental impact of incinerating discarded PVC gloves.

PVC gloves are manufactured by plastisol dip coating of cotton liners covering the moulds. For this process the rheology profile over the full range of processing temperatures is extremely critical. Incorrect formulation can give problems of excessive coating weight, plastisol drips or 'strike through' of the woven cotton liner. The number of commercial paste polymers suitable for this application is very limited, the most successful being Pevikon® 737 (Hydro Polymers).

For heavy duty chemically resistant gloves the plasticisers used are triaryl phosphates, not in this instance for their fire resistance but because of their resistance to chemical hydrolysis and to extraction. User specifications demand retention of properties after contact with a variety of aggressive liquids including concentrated mineral acids, oils and solvents. The critical rheological requirements of the dipping process differentiate between different types of triaryl phosphates and coal tar based trixylyl phosphate (TXP) is usually the preferred option.

Lighter weight general purpose gloves normally contain DOP as sole plasticiser. When it was available di-iso-octyl phthalate was preferred technically because experience had shown it to give fewer problems in the process – a conclusion not predictable from standard rheological data.

6.14 ADHESIVE SHEET AND TAPES

These are produced by adhesive coating of thin calendered sheet. The factors influencing the selection of plasticisers for these applications are discussed in Chapter 3 (3.6.3). Commodity phthalates are used in decorative sheet and labels. Avoidance of migration into the adhesives used in high temperature electrical tapes and medical tapes requires the use of polyester plasticisers.

Production of foamed sealing tapes for draft exclusion etc involves the coating of chemically blown PVC with adhesive. For products used at normal ambient temperatures a commodity phthalate is adequate as the plasticiser although a proportion of butyl benzyl phthalate may be included to enhance fusion and give optimum expansion and cell structure.

Industrial sealing tapes for heavier duty are required to be free of plasticiser migration either into adhesives at elevated operating temperature or into any polymeric surface in contact with the seal. Trimellitates (especially tri-2-ethylhexyl trimellitate) are in use for these more demanding applications. Non-migratory polyester plasticisers would be generally unsuitable because of their high plastisol viscosity.

6.15 CALENDERED PRODUCTS – MISCELLANEOUS

Calendered flexible PVC is used for many applications in addition to the major outlets in automotive interiors and construction which have been mentioned earlier. These include:

stationery goods – wallets and folders, clear and opaque
waterproof mattress covers
shower curtains
dust covers for office equipment etc.

There are niches where specifications may create a need for the use of speciality plasticisers but these are unusual and there is very little demand in these applications for plasticisers other than general purpose phthalates. There are preferences for DOP, DINP or DIDP in specific cases according to individual requirements relating to processing, end properties and costing.

In the main the environments encountered by these products are likely to cause only negligible plasticiser loss although interesting exception sometimes occur. Examples have been observed in babies' cot mattresses and changing mats where some areas of the covers were found to have stiffened as a result of having lost a large proportion of their plasticiser. This was most conventionally attributed to extraction by skin creams and oils. A more surprising explanation for which there was considerable evidence in some cases was chemical degradation of the PVC causing crosslinking with consequent reduction in plasticiser compatibility. Factors thought to be promoting degradation were migration of the triethylene diamine catalyst from the polyurethane foam[95] and absorption of urine.[96]

6.16 COATED FABRICS – MISCELLANEOUS

Fabric coating was one of the first applications to be developed for plasticised PVC. PVC leathercloth soon became accepted as a cheap and durable substitute for traditional materials in vehicle seats, furniture, footwear, fashion clothing, handbags, luggage ware and protective clothing. In turn the use of PVC has been eroded by alternative materials, for example as already mentioned textiles have taken over for the front surfaces of car seats. However the cost/performance of PVC coated fabrics ensures their continued use in a wide variety of outlets.

In most cases lowest cost plasticiser systems comprising any general purpose plasticiser (with or without a C_{15} 45% chlorinated paraffin) meet all processing and end product requirements. There are limited exceptions which may involve the use of a speciality plasticiser. For example fire resistant seating used in public transport may incorporate an alkylaryl phosphate; tractor seats for use in very cold climates have severe cold crack requirements necessitating the use of a low temperature plasticiser; linear phthalates have been used to confer good cold flex to industrial protective clothing.

In the early nineteen seventies at the time when PVC was popular for domestic upholstery there were regular occurrences of a phenomenon which became known as 'leathercloth hardening'. The causes of this were not immediately obvious. The effect occurred most typically in armrests where the coating was under tensile stress combined with regular frictional contact with the palms of the hands. Cracking and separation of the coating from the fabric was found to be associated with substantial plasticiser loss. A variety of potential contributory factors were postulated including the effects of cleaning media and even the possibility that certain individuals might be potent extractors or degraders of plasticisers! In some cases there was evidence that low temperature dehydrochlorination and crosslinking of PVC was playing a part, probably initiated by migration of triethylene diamine catalyst from the polyurethane upholstery foam.[95]

6.17   EXTRUDED PRODUCTS – MISCELLANEOUS

Published surveys of the flexible PVC market are often based on the type of conversion process and hence contain the broad category 'extrusions'. This excludes extrusion of cable compounds, the largest example, since it is a distinct sector with clearly defined characteristics.

Other applications of extruded plasticised PVC have been covered earlier in this chapter and the factors influencing plasticiser selection in specific cases have been discussed. The construction and automotive industries use soft extruded PVC products for sealing purposes. Profiles used in car interiors are subject to windscreen fogging specifications whilst those used for exterior trim need to be resistant to weathering. Clear PVC tubing is used in health care for conveyance of biological fluids.

This still leaves a variety of other end uses for flexible PVC extrusion compounds. Hose of various types for conveying air and aqueous liquids provides the main demand in this miscellaneous sector. A major proportion of this is served by compound producers supplying materials selected from their standard product ranges. In practice performance requirements seldom place technical demands on plasticisers which cannot be satisfied by commodity phthalates and the consumption of specialities is extremely small.

CHAPTER 7

Plasticisers for Polymers other than PVC

7.1 INTRODUCTION

Polymers other than PVC account for less than 10% of plasticiser consumption. Whilst there are many polymers in which plasticisers can perform a useful function the actual volumes used are in most cases very small. This chapter is not intended to be a comprehensive review of such applications and gives emphasis to those polymers where the use of plasticisers is of greatest commercial and technical importance at the present time.

The reasons for the almost unique capability of PVC to be tailored into a range of useful flexible materials by the addition of high plasticiser levels have been discussed in Chapter 1. Hardly any of the polymers discussed in this chapter share this characteristic and in general the levels of plasticiser addition are very much lower than with PVC. With PVC the main purpose of adding plasticiser is to convert the rigid polymer into a flexible thermoplastic material with physical properties suited to a specific end use. With other polymers the major function of the plasticiser is nearly always different, for example:

Polymer	*Plasticiser Improves:*
Polyvinyl acetate	Minimum film forming temperature of emulsions.
	Adhesive bond flexibility.
Cellulose acetate	Melt flow.
	Impact resistance.
Polyvinyl butyral	Adhesion to glass.
	Energy dissipation in glass laminates.
Nitrile rubber	Mouldability.
	Cold flex.
Polysulphide and polyurethane sealants	Viscosity.
	Cost.
Polyester engineering thermoplastics	Mechanical strength.

Once we depart from the sphere of PVC the classification of additives as plasticisers can become somewhat blurred. They could alternatively be described in specific cases as coalescing agents, process aids, impact improvers or crystallisation accelerators. Hence drawing parallels with PVC technology will not always necessarily be helpful.

The majority of the polymers covered in this chapter differ from PVC in one important respect – they are more expensive, sometimes by a factor of more than four. Together with the fact that plasticiser levels are usually much lower than in PVC this means that they contribute a much lower proportion of the end product cost. Hence there is more of a tendency than in the PVC industry to use relatively expensive speciality plasticisers.

7.2 POLYVINYL ACETATE (PVAc)

7.2.1 EXTERNAL VERSUS INTERNAL PLASTICISATION

PVAc is produced by emulsion polymerisation of vinyl acetate in the presence of suitable initiators and emulsion stabilising systems. The emulsions are used as bases for adhesives, textile and paper binders, sealants and paints. Although paints constitute the largest outlet for vinyl acetate polymer emulsions it is in adhesives that the majority of additive (external) plasticiser is consumed. The glass transition temperature of unplasticised PVAc homopolymer is in excess of 30°C which is too high for it to be used in paints and many adhesive applications. At lower temperatures particle coalescence and film formation are not achieved. Even when the temperature is high enough to allow film formation it may be too brittle to withstand substrate movement.

Nearly all PVAc paint is now based on internally plasticised copolymers in which the plasticising comonomer is either ethylene or vinyl versatate (VeoVA®), the second of these predominating.

It can be seen that there are two different structural forms giving internal plasticisation. The effect of ethylene is to provide flexible chain segments whereas VeoVA functions by creating free volume between chains as a result of its bulky side group.

PVAc copolymers incorporating ethylene are known as 'pressure polymers' because their production requires the use of a high pressure process to bring ethylene into the liquid phase for polymerisation. There are variations on this type of product in which vinyl chloride is included as a cheap antiplasticising monomer and/or a functional monomer is included to provide sites for crosslinking. Pressure polymers are used in large volumes in adhesives and ever since their introduction have been steadily displacing externally plasticised PVAc homopolymers. Despite the saving in raw material cost resulting from the replacement of vinyl acetate by ethylene, pressure polymer emulsions have generally remained significantly more expensive than homopolymers. This is a consequence of the high capital cost of the high pressure plant relative to the simple equipment used for the manufacture of homopolymer emulsions. Pressure polymer show significantly better adhesive performance than homopolymers. Setting speeds are faster and they are capable of bonding to a wider range of surfaces some of which were traditionally considered 'difficult' for emulsion adhesives (e.g. plastics in general). In addition they offer an escape from the health, safety and environmental questions which perpetually surround the use of external plasticisers.

7.2.2 EXTERNAL PLASTICISERS FOR PVAc.

At temperatures less than about 30°C unplasticised PVAc homopolymer is a relatively brittle material. In lowering its glass transition temperature in emulsion adhesives a plasticiser performs a dual function. Firstly it allows particle coalescence to occur as the adhesive film dries (usually by migration of water into the substrate). This leads to the formation of a coherent homogeneous material, that is to say in conventional terms it reduces the minimum film forming temperature (MFFT) of the adhesive. Secondly it gives a more flexible bond which is a requirement for many applications. The presence of the plasticiser provides better bonding to some less porous substrates such as clay coated papers.

Levels of addition of plasticiser to PVAc range up to about 15 p.h.r. (based on dry weight of polymer). Many common plasticisers have much higher limits of compatibility with the polymer and are capable of giving stable homogeneous plasticised compositions with very high levels of addition. However PVAc is amorphous and lacks the microcrystalline structure which constrains the deformation of plasticised PVC. Hence high levels of plasticisation give materials which are of little value because of their high creep.

Plasticisers are readily dispersible in PVAc emulsions by agitation at slightly elevated temperatures (typically 30°C). The plasticiser is essentially absorbed completely and uniformly into the polymer particles during this process. It has been shown that this is in fact a prerequisite for emulsion stability and efficient reduction of MFFT.[97]

Adhesives constitute the largest outlet for PVAc homopolymer emulsions accounting for about 60% of Western European demand in 1992. Roughly a further 20% was used in textile applications including non-woven products. The main application of plasticised PVAc adhesives are in the packaging industry where they are used for bonding paperboard boxes and cartons. Their dominance of this market is a result of their low price and ease of processing. As far as it is possible to determine for emulsion polymers, typical values of molecular weight for such adhesives are in the region of 500,000. Emulsions for wood adhesives are produced under conditions favouring higher molecular weight in order to reduce bond creep. In this application flexibility is not a requirement and the products are not plasticised. In the absence of plasticiser reduction of MFFT relies on the incorporation of a **coalescing solvent**. A coalescing solvent is a transient plasticiser which having facilitated film formation is lost from the composition through migration and evaporation. They are usually esters of relatively low molecular weight sometimes containing a hydroxy group. Texanol® (see Chapter 5, 5.11) is a well known example which for many years held a major market share. Factors influencing the design and selection of coalescing solvents have been reviewed recently by Hazell.[97]

The plasticisers which have long been the PVAc industry workhorses are the C_4 phthalates, di-n-butyl phthalate (DIBP) and di-isobutyl phthalate (DIBP). Because of their high interaction with the polymer and their low viscosity they are easily incorporated into emulsions. Their vapour pressure is sufficiently low for them to be considered essentially involatile in all but a few applications of PVAc. Above all, their advantage over any technical alternative has been their relatively low cost. DBP and DIBP are readily interchangeable although the somewhat lower efficiency of DIBP may necessitate some adjustment in addition levels.

Table 7.1 compares the performance of DBP and DIBP added at 10 p.h.r. (on the basis of dry weight) to a PVAc emulsion with solids content 55.5%. Figures 7.1 to 7.4 show the effect of increasing DBP level on MFFT of an emulsion with PVAc solids content 41.9 % and on the physical properties of the dried film.[98]

In the past many compounders of PVAc emulsions preferred DIBP to DBP because it offered significant cost saving. When the price differential was eroded by the gradual disappearance of surplus isobutanol (Chapter

TABLE 7.1
DBP VERSUS DIBP IN PVAc EMULSIONS.

PLASTICISER			None	DBP	DIBP
Viscosity	7 days	mPa.s	12.3	14.7	16.0
	14 days	mPa.s	13.4	14.9	15.8
MFFT		°C	14.8(*)	2.5	4.7
Dry film properties					
Pendulum Hardness (DIN 53 6469)			78	10	13
Tensile strength		MPa	8.8	3.7	5.1
Elongation at break		%	0	370	340
Volatility at 70°C,	1 day	%	0.1	0.7	1.0
	14 days	%	0.8	3.3	4.6

(*) base polymer presumed to have contained some coalescing solvent.

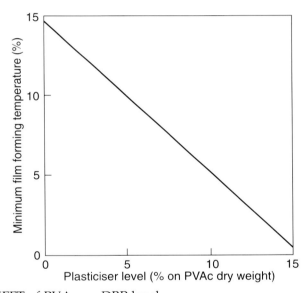

Fig 7.1 MFFT of PVAc vs. DBP level.

4, 4.5.2) the majority switched to DBP without any problem. However, at the time of writing events are occurring which has caused considerable upset in this market. These are a have been set in train by new European legislation governing the labelling of containers containing hazardous substances. Containers holding either DBP or 'preparations' (e.g. adhesives) incorporating more than 5% of DBP are required to be labelled with certain 'risk phrases'. The situation is complex and fraught with misunderstanding as discussed in detail in Chapter 9. However the main points to note here are as follows:

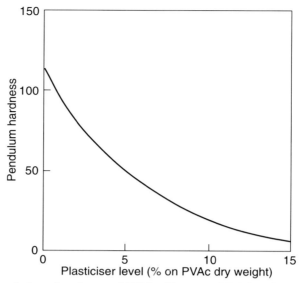

Fig 7.2   Pendulum hardness of PVAc films vs. DBP level.

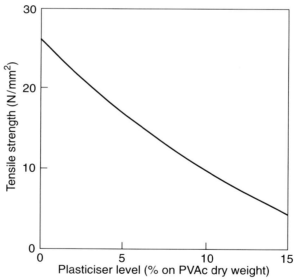

Fig 7.3   Tensile strength of PVAc films vs. DBP level.

(a) Containers of adhesive containing more than 5% of DBP are subject to the regulations. This includes a very high proportion of formulations in common use.

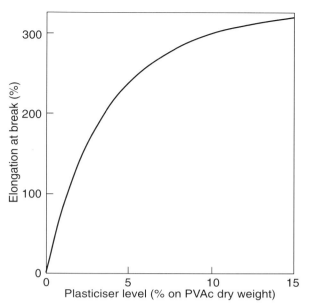

Fig 7.4 Elongation at break of PVAc films vs. DBP level.

(b) Although the nature of the 'risk phrases' inevitably causes concern among downstream users of the products their real significance in relation to risk is little understood.

(c) At present there is no requirement to label DIBP. Basically this is because for DIBP there is a relative lack of information which would allow clear categorisation according to the rules. Hence for the present DIBP is differentiated from DBP and in this respect offers a short term escape from labelling requirements. Whilst this differentiation appears to have precipitated a swing back to DIBP it is recognised by most informed parties that the situation is artificial and unlikely to last.

(d) Exercises in DBP avoidance may well result in substitution by plasticisers other than phthalates which usually escape classification for the main reason that they have been relatively little studied. The probable long term consequence of this situation is an acceleration of the replacement of homopolymers by internally plasticised copolymers.

A study of plasticiser suppliers' literature indicates PVAc as an outlet for a wide range of plasticiser types including triacetin, benzoates, citrates, polyesters and chlorparaffins and even sulphonamides. For compatibility with PVAc these materials typically have higher solubility parameters than general purpose PVC plasticisers. In practice the use of plasticisers other than C_4 phthalates has been confined to small niche applications. In such cases they might be needed for their higher permanence or conformance to toxicological approvals. Table 7.2 compares the performance of DBP with that of a short chain polyester plasticiser specially designed for use in PVAc binder and adhesive applications involving elevated process temperatures and contact with non-aqueous extracting media.[99] Tests were carried out on cast films containing 16.5 p.h.r. of plasticiser (on polymer dry weight).

TABLE 7.2
Comparison of PVAc plasticised with DBP and a polyester plasticiser

		DBP	Reoplex® 400
Plasticiser viscosity at 25°C	mm²/s	16	7000
Cold crack temperature	°C	−1	+5
Sward rocker hardness		6	8
Volatile loss	%	9.1	1.2
Extraction by: petrol	%	1.6	0.4
mineral oil	%	1.2	0.4
olive oil	%	1.3	0.5
water	%	4.0	4.0
soap solution	%	4.5	14.0
detergent solution	%	4.0	9.0
perchlorethylene	%	3.0	1.8

A 56% chlorinate of C_{10-12} paraffins was at one time promoted as a general substitute for DBP but the relatively small cost saving which it gave was seen as insufficient to compensate for its poorer processing characteristics. In the USA benzoates (particularly dipropyleneglycol dibenzoate) have become widely used as replacements for DBP as a result of users wishing to distance themselves from the concerns associated with safety of use of phthalates. Until recently any such tendency in Europe has been inhibited by relatively high prices for benzoates.

7.3   POLYVINYL BUTYRAL (PVB)

Plasticised polyvinyl butyral is used as the adhesive interlayer in laminated safety glass for automotive and architectural applications. Despite being used in this application for more than 50 years it still has no major competitors and continues to show high growth.[100] PVB is also used to a much lesser extent in coating applications. Although plasticiser consumption with PVB is far higher than with most other polymers it has tended to receive only passing attention in this type of review. This is probably due to the nature of the market which involves only a very small number of suppliers and customers at each stage of the sequence.

PVB RESIN ⟍

⟶ EXTRUDED PVB SHEET ⟶ GLASS LAMINATE

PLASTICISER ⟋
(and other additives)

The largest players in this market, Monsanto and DuPont operate globally and manufacture both PVB resin and the plasticised sheet. Selection of plasticisers is based on lengthy development programmes by the sheet manufacturers. Purchasing specifications are highly defined and relate to the severe demands placed on the end product. The glass laminates are required to maintain their integrity over many years of service without any colour development, loss of transparency or decline in impact performance. Plasticiser supplies are procured by sheet manufacturers with no applications technology input from the suppliers.

PVB is a derivative of vinyl acetate via the following sequence.

vinyl acetate ⟶ poly(vinyl acetate) ⟶ poly(vinyl alcohol) ⟶ PVB
HYDROLYSIS   CONDENSATION WITH
                BUTYRALDEHYDE

In the grades of PVB used for glass laminates about 20% of the hydroxyl groups of the polyvinyl alcohol remain unsubstituted.

$$-CH_2-CH-CH_2-CH-CH_2-CH- \quad \text{polyvinyl alcohol}$$
$$\qquad\; | \qquad\qquad | \qquad\qquad |$$
$$\qquad OH \qquad\quad OH \qquad\quad OH$$

$$\downarrow \quad CH_3-CH_2-CH_2-CHO \quad \text{butyraldehyde}$$

$$-CH_2-CH-CH_2-CH-CH_2-CH-$$
$$\qquad\; | \qquad\qquad | \qquad\qquad |$$
$$\qquad O \qquad\quad\; O \qquad\quad OH$$
$$\qquad\;\; \diagdown \qquad \diagup$$
$$\qquad\qquad CH$$
$$\qquad\qquad |$$
$$\qquad\qquad CH_2$$
$$\qquad\qquad |$$
$$\qquad\qquad CH_2$$
$$\qquad\qquad |$$
$$\qquad\qquad CH_3$$

The polymers so produced have high clarity and show remarkable adhesion to glass. In the absence of plasticiser PVB is a brittle material at ambient temperature. By the use of a longer chain aldehyde (e.g. C_{10}) as a partial substitute for butyraldehyde it is possible to introduce a degree of internal plasticisation thereby allowing a reduction in the level of external plasticiser.[101] However at present external plasticisation is the norm.

Compounds containing resin, plasticiser (typically at a level of about 40 p.h.r.) and the various minor additives are produced by hot melt processing and extruded into sheets of the required thickness (generally 0.25 mm for automotive and 0.17 mm for architectural glass). Measures are required to prevent blocking of the rolled sheet prior to lamination. Nowadays this is usually achieved by the use of refrigerated transport and storage (in the range 2–8°C). Polyethylene interlayers are sometimes used as an alternative.

Whilst very many plasticisers are compatible with PVB the few used in practice in safety glass applications are all aliphatic diesters. At one time the common industry standard was **triethyleneglycol di-2-ethylbutyrate**, best known as Flexol® 3-GH.

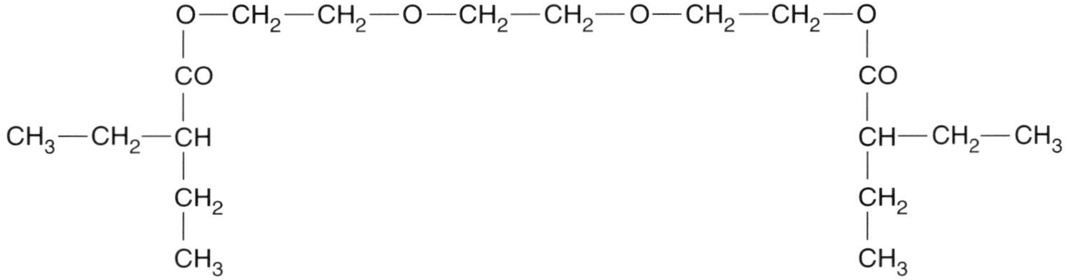

$$O-CH_2-CH_2-O-CH_2-CH_2-O-CH_2-CH_2-O$$

triethyleneglycol di-2-ethylbutyrate
Molecular weight 346

The other plasticiser commonly mentioned in older texts on the subject is **dibutyl sebacate**, not dissimilar to the previously mentioned ester but lacking its ether groups.

A period of supply problems with the 2-ethylbutyrate in the nineteen seventies forced the PVB sheet producers to find suitable alternatives. Prominent among the plasticisers used currently are **tetra(ethyleneglycol) di-n-heptanoate** and **di-n-hexyl adipate**. The latter is isomeric with the more expensive dibutyl sebacate.

Whilst it is possible that alternative plasticisers might be found which gave cost/performance advantages over those mentioned above this is one market where the need for continuity is likely to be over-riding. Hence changes in the foreseeable future are unlikely.

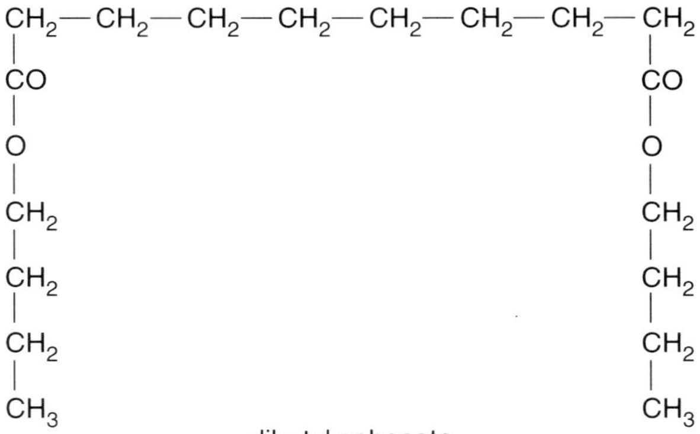

$$CH_2\!-\!CH_2\!-\!CH_2\!-\!CH_2\!-\!CH_2\!-\!CH_2\!-\!CH_2\!-\!CH_2$$

```
CH₂—CH₂—CH₂—CH₂—CH₂—CH₂—CH₂—CH₂
|                                   |
CO                                  CO
|                                   |
O                                   O
|                                   |
CH₂                                 CH₂
|                                   |
CH₂                                 CH₂
|                                   |
CH₂                                 CH₂
|                                   |
CH₃                                 CH₃
```

dibutyl sebacate
Molecular weight 314

```
O—CH₂—CH₂—O—CH₂—CH₂—O—CH₂—CH₂—O—CH₂—CH₂—O
|                                              |
CO                                             CO
|                                              |
C₆H₁₃                                          C₆H₁₃
```

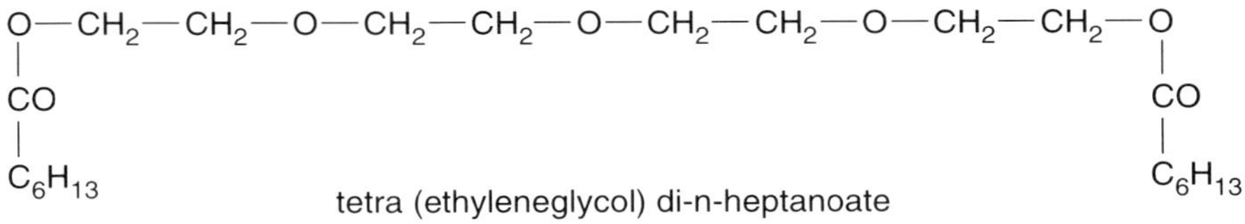

tetra (ethyleneglycol) di-n-heptanoate
Molecular weight 418

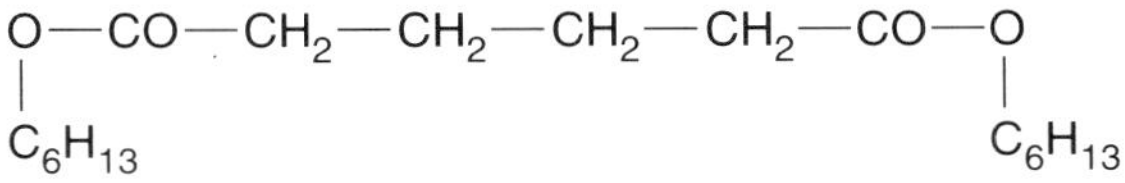

```
O—CO—CH₂—CH₂—CH₂—CH₂—CO—O
|                        |
C₆H₁₃                    C₆H₁₃
```

di-n-hexyl adipate
Molecular weight 314

7.4   CELLULOSICS

7.4.1   GENERAL

Cellulose based materials are the oldest of all thermoplastics and date from the nineteenth century. As related in any history of plasticisers these materials were the cradle of plasticiser technology. Cellulosic plastics are mainly derivatives of cellulose in which an average of two out of the three hydroxyl groups in each structural repeat unit have been esterified. Cellulose has the structure shown on the following page.

This can be expressed as $(C_6H_{10}O_5)_n$ where n is a number in excess of 3000. During the conversion processes chain scission occurs resulting in very much shorter sequences in the ester derivatives (n = 175–360 in commercial acetates).

7.4.2 CELLULOSE ACETATE

Cellulose acetates are produced by the reaction between cellulose and acetic anhydride in the presence of sulphuric acid. This gives the completely acetylated product cellulose triacetate which can then be partially hydrolysed to give a range of useful materials. In commercial products the degree of OH substitution ranges from 2.2 to 3.0 (acetyl content 36.5 to 44%). The materials used for injection moulding compounds, clear 'acetate' sheet and cigarette filters are essentially diacetate. The triacetate is used for production of fibres (best known as Tricel®) and some specialised types of film.

Not surprisingly in view of its advanced maturity cellulose acetate has been supplanted by more modern materials for many applications. The number of producers of cellulose acetate moulding compositions in Europe has declined to three or four. However the characteristic gloss, clarity and toughness of the material keeps it in widespread use for the moulding of items such as tool handles, spectacle frames, toothbrushes and hair combs. Clear extruded sheet is used as a high gloss cover coat for laminating to paper, for clear windows in packaging cartons and for 'acetate' transparencies.

Unplasticised cellulose diacetate has a tendency to decompose at temperatures below its softening point and compositions for injection moulding and extrusion contain up to about 50 p.h.r. of a plasticiser selected from a small group of commercial options. In addition to facilitating processing the plasticiser gives beneficial modification of end properties – increased flexibility in sheet applications and better impact resistance in mouldings. The workhorse of the plasticisers used by the industry is **diethyl phthalate**, DEP. The more volatile **dimethyl phthalate**, DMP at one time held this position but it is now reserved for meeting special processing requirements where its greater efficiency in reducing melt viscosity are exploited. Table 7.3[102] shows the effect of various levels of DMP on the properties of cellulose diacetate.

TABLE 7.3
PROPERTIES OF CELLULOSE DIACETATE PLASTICISED WITH DMP

p.h.r. DMP		22.6	30.0	37.8
Flow temperature	°C	153	140	130
Elongation at break	%	5.0	6.5	8
Tensile strength	MPa	68.4	55.9	45.7
Rockwell hardness		101	90	79
Water absorption	%	3.2	2.5	2.1
Leaching, 48 hours	%	0.45	0.52	0.58
Loss in weight after 1 week at 150°C (*)	%	0.53	0.86	2.4

(*) These losses are surprisingly low in view of the apparent severity of the test conditions. However details of the test used here are unavailable.

For compatibility with cellulose acetate plasticisers need to be much more polar than those used with PVC. Those used in practice are mainly materials with relatively low molecular weight and correspondingly high vapour pressure. These long established plasticisers continue to be quite acceptable for most applications and processes and there appears to be little pressure to move to higher molecular weight materials to reduce volatile loss. For applications where the permanence of DEP is inadequate formulators have traditionally resorted to **dimethoxyethyl phthalate** (dimethylgycol phthalate, DMGP) or **dibutyl phthalate** (DBP) although the latter has a significantly lower compatibility limit than DEP. Low molecular weight polyesters have limited used with cellulose diacetate in niche applications requiring high permanence. **Triphenyl phosphate** (TPP), a solid with a melting point of 48.5°C, is effective in reducing melt viscosity but is inefficient in lowering hardness. It is used where a hard surface and/or enhanced fire resistance are required.

The specialised large scale use of **triacetin** in the production of cigarette filters from cellulose diacetate fibre is described in Chapter 5, 5.12. On a much smaller scale DEP is used as a carrier in the disperse dyeing of Tricel fibre. In this application its function is to facilitate penetration of the fibre surface by a non-reactive dye.

7.4.3 CELLULOSE NITRATE ('NITROCELLULOSE')

Cellulose nitrate plasticised with the natural product camphor has an honoured place in materials technology as 'Celluloid' the first (partly) synthetic thermoplastic.

camphor

Whilst the level of incorporation of camphor into this material was quite high (typically 50 p.h.r.) its primary function as a plasticiser was to reduce melt viscosity rather than to confer flexibility to the end product. It is generally considered that technology of plasticisers dates from the search for synthetic substitutes for camphor for this application in the early part of the twentieth century. During the intervening period melt processing of nitrocellulose has declined to a very low level and it now survives in a few specialised niches.

In solution form nitrocellulose remains an important material for a number of applications the foremost of which is as a pigment binder for flexographic and photogravure inks used by the packaging industry. 65–70% of packaging ink used currently contains nitrocellulose. Its continuing dominance in this market results from its combination of good solvent release, low odour, heat resistance, good pigment wetting and relatively low cost. It is also used in inks for the silk screen printing of hoardings. Nitrocellulose lacquers were once the main type of car finishes but were superseded by more durable acrylics. However they remain in small scale use for some types of car refinish. Nitrocellulose lacquers are still widely used for wood finishing. A further specialised application is heat sealable coatings for regenerated cellulose film (Cellophane).

In all of the various applications of nitrocellulose it is normal practice to include a plasticiser to increase the flexibility and impact strength of the dried composition. Although cellulose dinitrate has a high degree of crystallinity it readily accepts high proportions of plasticiser because its melting point is easily depressed. Unlike cellulose diacetate it is tolerant of a wide range of plasticiser polarity. DOP is compatible with the dinitrate at a level of 100 p.h.r whereas it shows very limited compatibility with the diacetate. Levels of addition in practice are much lower than this, typically being around 30 p.h.r. The best all round cost/performance is shown by **dibutyl phthalate** which has been by far the most widely used plasticiser for nitrocellulose inks and lacquers. Small quantities of other plasticisers are used for special requirements, for example phosphates for fire resistance or dicyclohexyl phthalate for blocking resistance.

7.4.4 CELLULOSE ACETATE BUTYRATE AND CELLULOSE PROPIONATE

The molecular structure of these materials contain larger side chains than those present in cellulose diacetate. These provide a degree of internal plasticisation which allows the use of lower levels of external plasticiser to achieve the same processing and end properties. This can be of advantage when questions of plasticiser permanence arise. These materials are manufactured by Eastman Chemical Company in the USA where they have more of a differentiated market position relative to (plasticised) cellulose diacetate than in Europe.

7.5 RUBBERS

7.5.1 PROCESS OILS

There has always been something of a divide between the technology of thermoplastics and the older established rubber technology which started its evolution long before the era of synthetic materials (Charles Goodyear discovered the process of vulcanisation in 1838). Differences in terminology can sometimes hinder communication between technologists on either side of the divide when in reality they are talking about very similar things. The subject of plasticisers provides an example.

The rubber industry uses large volumes of 'process oils' or 'extenders' in tyres and general rubber goods manufactured from the hydrocarbon rubbers. Consumption in the UK in 1993 was estimated at 60–70,000 tonnes. These oils which are refinery products are classified by the industry as 'aromatic', 'naphthenic' (saturated cyclic structures) or 'paraffinic' although since they are complex mixtures distinction between the three types is not precise. Their functions in rubber compounds are:

(a)   to facilitate processing
(b)   to increase the softness, extensibility and flexibility of the end product
(c)   to reduce cost.

They are normally used in compositions containing high levels of filler to compensate for their negative effects on processabilty and softness. Process oils could reasonably be described as plasticisers although the rubber industry tends to reserve this term for the more specialised additives described in 7.5.2.

Selection of oil type is influenced mainly by considerations of compatibility with different rubber structures. Levels of addition can range from about 10 p.h.r. to 100 p.h.r. Concerns over the toxicity of the low priced aromatic oils are being addressed by measures to remove the hazardous components from these products. The scope for direct substitution by non-aromatic oils is constrained by their compatibility limits. For some high performance applications requiring good heat and light stability highly refined 'white oils' with prices 3–4× those of aromatics are used.

Hydrocarbon process oils for rubbers have been disregarded in statements made in this book on relative consumption of different types of plasticiser.

7.5.2 Synthetic plasticisers for rubbers

Hydrocarbon process oils are unsuitable for use in synthetic rubbers with polar molecular structures partly because of their limited compatibility. In their place synthetic materials containing polar groups, most commonly esters, are used. With these polymers the principal function of such additives is usually to extend the operating temperature range of the end product by increasing low temperature flexibility without adversely affecting high temperature ageing resistance. For rubber products used in contact with oily media, plasticisers are chosen so that the opposing effects of extraction and swelling are balanced.

The majority of demand for plasticisers in the rubber industry is for polychloroprene (CR) and nitrile (NBR) rubbers which are based on the monomers:

FOR CR	$CH_2=CH-CCl=CH_2$	chloroprene
FOR NBR	$CH_2=CH-C=N$	acrylonitrile
and	$CH_2=CH-CH=CH_2$	butadiene

NBR is used principally for automotive applications (seals etc.). CR is used for automotive components, conveyor belts and a variety of other applications. It is estimated that the consumption of plasticisers with these polymers in Europe is of the order of 10kt/a. There is a parallel with the PVC industry in that the majority, perhaps 80% of this volume is accounted for by general purpose phthalate plasticisers, predominantly

DOP together with DINP and DIDP. Some rubber compounders are also producers of PVC compounds and use common bulk storage of phthalates for both outlets. As with PVC, the same groups of speciality plasticisers are used to meet special requirements – low temperature flex, high temperature ageing resistance, fire resistance or migration resistance. Published data on their performance in rubbers is very much more sparse than with PVC. Consequently PVC data are commonly used to predict the **relative** performance of different plasticisers in rubbers. Typical levels of addition of plasticisers to rubber are less than 20 p.h.r. which is much lower than with PVC. Hence the degree of modification of physical properties of the polymer is less than with PVC.

The rubber industry is notable for the diversity of plasticisers which it uses in its myriad of end products. In this sense rubbers provide a complete contrast to PVB covered earlier in this chapter. The number of rubber processors in Europe is large. Many of them manufacture extensive ranges of products using a wide variety of formulations developed over a long period of time. In many cases it can be argued that there is scope for rationalisation in the use of plasticisers with consequent savings in material costs. However the development costs of reformulation often preclude this for the small volumes involved and so the diversity persists.

The majority of speciality plasticisers used with rubbers have structures with a high degree of linearity which favours good low temperature performance. A recent survey by one supplier listed more than 50 such products commercially available in Europe. Differences in molecular weight and polar group content provide them with a wide range of polymer compatibility, volatility and resistance to extraction. Some of them are familiar to PVC formulators but others are unknown outside the rubber industry. Occasionally products marketed under trade names with only a vague disclosure of composition are in fact relatively common materials or blends of such. The following list indicates the variety of low temperature plasticisers currently used by the European rubber industry.

Dialkyl adipates	di-2-ethylhexyl adipate	DOA
	di-isodecyl adipate	DIDA
	mixed adipates, C_8–C_{10}	
Dialkyl azelates	di-2-ethylhexyl azelate	DOZ
	di-isobutyl azelate	
Dialkyl sebacates	di-n-butyl sebacate	DBS
	di-2-ethylhexyl sebacate	DOS
Ether adipates/	di(butoxyethoxyethyl)adipate	BCA
glutarates	adipate of triethoxybutanol	
	'dialkyl diether glutarate'	

Polyethers	butyl Carbitol formal	BCF
Glycol esters	triethylene glycol dihexanoate	
	triethyleneglycol diheptanoate	
	triethyleneglycol di-2-ethylhexanoate	
	triethleneglycol di-caprate/caproate	
Thioesters and thioethers	dibutyl methylene bis(thioglycollate)	
	di-2-ethylhexyl thiodiglycollate	
	'ether thioether'	

Di(butoxyethoxyethyl) formal (butyl Carbitol formal, BCF) shows outstanding low temperature performance and in some countries has long had the status of an industry standard in this respect. However it does have the disadvantages of greater water sensitivity and higher volatility than most of the products listed above. This is partly due to the fact that commercial products contain a significant proportion of the unconverted raw material butoxyethoxyethanol.

$$\textbf{BCF} \quad \begin{array}{l} O\text{-}C_2H_4\text{-}O\text{-}C_2H_4\text{-}O\text{-}C_4H_9 \\ | \\ CH_2 \\ | \\ O\text{-}C_2H_4\text{-}O\text{-}C_2H_4\text{-}O\text{-}C_4H_9 \end{array} \qquad \text{molecular weight 336}$$

A related plasticiser which has been found to be particularly versatile is di(butoxyethoxyethyl)adipate also known as butyl Carbitol adipate, BCA and butyldiglycol adipate.

$$\textbf{BCA} \quad \begin{array}{l} O\text{-}C_2H_4\text{-}O\text{-}C_2H_4\text{-}O\text{-}C_4H_9 \\ | \\ CO \\ | \\ (CH_2)_4 \\ | \\ CO \\ | \\ O\text{-}C_2H_4\text{-}O\text{-}C_2H_4\text{-}O\text{-}C_4H_9 \end{array} \qquad \text{molecular weight 434}$$

BCA has a useful level of compatibility with a wide range of rubber types in addition to NBR and CR and has been included in formulations based on hydrogenated rubber (HNBR), acrylic rubber (ACM), epichlorhydrin rubber, Vamac® and Hypalon®. Its fairly high molecular weight provides good heat ageing resistance at moderate test temperatures. The following tables show the performance of BCA relative to DOP in NBR and CR.[103] Most of the performance parameters shown are closely akin to those used for characterising plasticised PVC. Mooney viscosity indicates relative ease of processing of the composition prior to crosslinking.

TABLE 7.4
COMPARISON OF BCA AND DOP IN NITRILE RUBBER

PLASTICISER		None	BCA	DOP
Mooney viscosity 100°C	ML1+4	156	65	71
Hardness	IRHD	84	74	75
Tensile strength	MPa	23.2	18.2	17.7
200% modulus	MPa	22.7	15.4	14.6
Elongation at break	%	205	260	275
Gehman T10				
(cold flex temp.,BS 903 A13)	°C	−19	−33	−23
Compression set				
(70 hours, −5°C)	%	9.4	5.6	7.9
Heat ageing, 7 days 100°C				
Retained tensile strength	%	106	100	106
Retained elongation at break	%	86	87	80
Volatile loss	%	1.8	2.4	1.9
Hardness change		+4	+4	+4

Formulation:		
	*Breon® N33C50	100
	Zinc oxide	5
	N550 FEF Black	80
	Aminox	1.5
	ZMBI	1.5
	CBS	3.0
	TMT	1.5
	TET	1.5
	MC Sulphur	0.5
	Stearic acid	1.0
	PLASTICISER	20

* 50 Mooney, 33% acrylonitrile polymer with good processing properties

TABLE 7.5
COMPARISON OF BCA AND DOP IN POLYCHLOROPRENE RUBBER

PLASTICISER		None	BCA	DOP
Mooney viscosity, 100°C	ML1+4	62	29	39
Hardness	IRDH	66	68	71
Tensile strength	MPa	17.2	11.0	11.4
100% modulus	MPa	1.7	2.4	2.9
Elongation at break	%	410	250	260
Brittleness temperature (ASTM D2137–83)	°C	−32	−48	−39
Heat ageing, 70 hours at 100°C				
Retained elongation at break	%	83	92	83
Volatile loss	%	0	1.0	2.8
Hardness change		+3	+3	+5
Final brittleness temperature (ASTM D2137–83)	°C	−34	−45	−35

Formulation:	Butaclor® MC323	65
	Butaclor MC122	35
	Stearic acid	0.5
	Magnesium oxide	4.0
	Permanox OD	2.0
	Permanox IPPD	1
	Permanox 6PPD	1
	Polyethylene wax	3.0
	GPF Black	55
	Hard clay	20
	Zinc oxide	5.0
	ETU	1.0
	TMTD	0.5
	MBTS	0.25
	PLASTICISER	30

As with PVC trimellitate esters are now well established as plasticisers for rubber compounds formulated for use at high temperatures. Both tri-2 ethylhexyl trimellitate (trioctyl trimellitate, TOTM) and linear C8/10 trimellitate have been used in Hypalon® (chlorosulphonated polyethylene rubber) compounds for high temperature-, oil- and fire-resistant (HOFR) cables.

7.6 ACRYLICS

'Acrylics' is a generic term covering homo- and copolymers of esters of acrylic and methacrylic acids:

$$CH2=CH-CO-O-R$$
acrylates

$$CH_2=\overset{CH_3}{\underset{|}{C}}-CO-O-R$$
methacrylates

There is considerable scope for internal plasticisation of these polymers by variation of R. As regards external plasticisation, whilst acrylics may be capable of accepting high loadings of additive plasticisers their use with these polymers is on a relatively small scale at present.

Dibutyl phthalate, DBP
Dibutyl phthalate has long been used as a minor component of poly-(methyl methacrylate) (PMMA) cast sheet. However its function here is not as a plasticiser but simply as a polymer-compatible carrier for the additives which are incorporated into the prepolymer syrup during the casting operation. The addition level (<1%) is too low to have any adverse effect on the rigidity of the final product.

The use of water based acrylic sealants has grown rapidly in the last few years particularly in the consumer sector. Relatively polar plasticisers such as **BBP** or the cheaper alternative **DBP**, are used at levels between 5 and 10 p.h.r. They have the combined functions of improving rheology, aiding coalescence and enhancing flexibility.

The anti-PVC campaign of recent years has prompted research into substitute materials in most application areas. This has boosted the development of plastisol grade acrylic polymers. They have a particle form which allows them to be dispersed in plasticisers to form plastisols with rheological and fusion characteristics similar to those of conventional PVC plastisols. Such materials were originally developed as specialised adhesives for application to oily steel. One recent patent application[104] refers to a polymer based on methyl methacrylate 68%, butyl methacrylate 30% and acrylic acid 2%. A plastisol is formed by dispersing this in 80 p.h.r. of **di-isononyl phthalate**. The main market targeted for acrylic plastisols is automotive sealants and underbody protection. Certain car constructors have declared the aim of phasing out PVC in favour of less contentious substitutes although attention has been focused on interior trim rather than underbody protection. There would be a considerable cost penalty in the use of acrylic underseal and to mitigate this the use of existing commodity plasticisers like DINP is desirable.

So far acrylic plastisols remain largely a contingency 'non-PVC' option available from sealant suppliers.

7.7 POLYSULPHIDES

Liquid polysulphide polymers are widely used as the basis of sealants for the construction and automotive industries. They are among the oldest of synthetic sealing materials having been commercialised in 1929. They can be cured in place at room temperature to give a solid rubber seal without shrinkage. The cured rubber has a very broad temperature range of operation and is resistant to weathering and solvent extraction. In some important areas, particularly automotive glazing, polysulphides have lost market share to newer polyurethane sealants.

Chemically the polysulphides are polymers of bis-(ethylene oxy) methane which contain disulphide linkages. The short polymer sequences are terminated with reactive mercaptan groups which provide sites for curing.

$$- - - - S - S - CH_2 - CH_2 - O - CH_2 - O - CH_2 - CH_2 - S - H$$

Plasticisers are incorporated into the liquid polymer at about 30 p.h.r. to reduce viscosity and to enhance the flexibility of the cured seal. Because of the high price of the polymer they also confer the benefit of reduced formulation cost.

The plasticisers used must be fully miscible with the liquid prepolymer and free of exudation from the cured rubber. Because of the high polarity of the polysulphide structure general purpose dialkyl phthalate plasticisers such as DOP are insufficiently compatible to be useful here. More polar materials such as ether esters and alkyl benzyl phthalates have the required degree of compatibility and it is the latter which are commercially established as plasticisers for polysulphides

Butyl benzyl phthalate, BBP
Butyl benzyl phthalate has adequate performance for some applications but for automotive glazing and sealed double glazing units there is a requirement for materials of low vapour pressure in order to avoid 'fogging' of the glass. In these cases octyl benzyl phthalate and Texanol® benzyl phthalate (Santicizer® 278 – see Chapter 4, 4.9.2) are commonly used.

7.8 POLYURETHANES

Together with silicones polyurethanes are now the dominant type of material in the sealants market which is centred on the construction and automotive industries. As in polysulphides the plasticiser in a PU sealant functions as a viscosity cutter. In addition it lowers the cost of the composition by virtue of being cheaper than the prepolymer and/or allowing the use of higher filler levels. Levels of incorporation of plasticisers in PU sealants are typically in the range 10–30% Higher phthalates, in particular **DINP** and **DIDP**, are generally suitable for most applications. Significant quantities of **Mesamoll** (see Chapter 5, 5.13) are also used in some types of PU sealant formulations.

Whilst the largest consumer of PU sealants is the building industry an outlet which has seen particularly rapid growth in recent years is direct glazing systems for car construction. Here a moisture curable PU sealant is applied to the windscreen frame which has been pre-moistened with a water spray. The polyurethanes used for water curable sealants are based on polyesters rather than hydroxy terminated polybutadiene which is commonly used as a starting material for other types of PU sealing systems. PU automotive glazing sealants have created significant new demand for plasticisers, in the order of 2kt/a being used in Europe for this application by a small number of sealant manufacturers. Whichever plasticiser is used it is required to conform to particularly stringent specifications limiting water content in order to avoid the risk of premature partial cure.

7.9   MISCELLANEOUS THERMOPLASTICS

Polyvinylidene chloride (PVdC) film for packaging is produced by extrusion of a vinylidene chloride/vinyl chloride copolymer containing a low level of plasticiser (around 5 p.h.r.) which is included as a process aid rather than an end property modifier. **Dibutyl sebacate** was formerly well established in this application but has been superseded to a large extent by **acetyl tributyl citrate**. Selection of the latter was made in the USA for reasons of food contact approval. The plasticiser content of the PVdC film used in Europe each year amounts to several hundred tonnes.

Although there are many references in scientific and trade literature to plasticisers in polymers other than those mentioned so far in this book the scale of use is relatively insignificant. In engineering thermoplastics low levels of plasticisers may be used to lower melting point in order to facilitate processing. They may be used in conjunction with nucleating agents to aid crystallization thereby increasing rigidity and tensile strength. In these applications increases in softness or flexibility are obviously not desired effects. The additives used here are not generally of the type familiar as PVC plasticisers.

7.10   THERMOSETS

Plasticisers are commonly included at a level of about 1 p.h.r. in phenol/ formaldehyde and urea/formaldehyde moulding powders. By improving flow properties they allow the use of more highly crosslinked resins thereby minimising mould shrinkage. In the case of phenolic resins the ubiquitous **dibutyl phthalate, DBP** has been used for this purpose. DBP and **triaryl phosphates** have been used at somewhat higher levels (ca. 10 p.h.r.) in cresol/formaldehyde paper impregnating resins for producing electrical grade laminates. Here their purpose is to decrease the brittleness of the laminate to facilitate cold punching.

Low levels of plasticisers are conventionally used with some other highly crosslinked resins, notably epoxies. Their function in such cases is to aid processing in some way (e.g. by viscosity reduction) rather than to modify end properties.

CHAPTER 8

Plasticiser Quality, Specifications and Analysis

8.1 TOTAL QUALITY MANAGEMENT

In this chapter we move to a subject where concepts have changed considerably in Europe within the last decade. At one time quality was generally seen as being associated with sophistication, exclusiveness and a complex specification. However after much re-education of both manufacturing and service organisations quality is now much more equated with a supplier's consistency in conforming to customers' requirements. Once the details of a product specification have been agreed between the two parties, regardless of how tight or loose, then the supplier is expected to ensure that every delivery conforms to that specification.

Of course product conformance is only one item of a package of requirements by which the customer judges supplier quality, others being communication, documentation, delivery timing and so on. Some major companies now carry out rigorous audits on suppliers as a basis for establishing long term supply relationships. Sometimes the questioning can extend beyond the requirements mentioned above and seek details of suppliers' underlying quality management strategies and systems.

These developments have to be viewed in the context of Total Quality Management (TQM). Most people working in progressive industrial companies have now heard of this and many are able to expound its principles although few would be able to attempt a concise definition of TQM. Many companies in Europe have embarked on major TQM initiatives as parts of their strategies for ensuring long term competitiveness. These initiatives have enlisted the ideas and assistance of quality gurus among whom the most eminent is the late William Edwards Deming. As is often told, Deming's ideas were enthusiastically adopted in Japan long before they had any currency in either Europe or his native USA. They are considered to have made a major contribution to Japan's industrial success.

Having decided in the nineteen eighties to adopt a TQM culture, some European companies then purchased off the shelf systems in order to expedite progress. Others developed their own in-house training pro-grammes. It is known that among British companies who started off enthusiastically down the TQM path there is disappointment that the benefits have not lived up to expectations.[105] As a result there has been considerable reappraisal of the process and a withdrawal from the evangelical approach of the early days. It has been argued that a major limitation on the success of TQM initiatives has been the failure to pay attention to the psychology of individuals' motivation, a factor stressed in Deming's later writing.[106]

8.2 QUALITY MANAGEMENT SYSTEMS

Notwithstanding the difficulties which have been experienced, TQM culture has had a considerable influence on the expectations which many customers now have of their suppliers. A tangible element of the move-ment towards quality improvement has been the widespread adoption of quality management systems. Some companies have operated well devel-oped systems of their own for many years. However, in the nineteen eighties the appearance of the British Standard BS 5750 and its adoption as the ISO 9000 set of standards provided for the first time a common basis for dealings on quality matters between organisations. The first companies to gain registration of their operations to this standard acquired a certain amount of prestige which was exploited for marketing purposes. Registra-tion has now become so much the norm in many industries that failure to have achieved this status will often exclude a company from consideration as a supplier. BS 5750, Parts 1, 2 and 3 were in 1994 subjected to minor revision and reissued as BS EN ISO 9001/9002/9003. A quality manage-ment system conforming to BS EN ISO 9002 can be used for control of the sequence from raw materials through manufacturing to despatch and delivery of the finished product to the customer. Registration is obtained through an independent assessment by an accredited certification body such as the BSI.

A key requirement for registration to the standard is the existence of quality manuals detailing all relevant procedures which will ensure that the customer receives product conforming to the agreed specification. These procedures must ensure that sufficient testing is carried out to guarantee that every delivery conforms to all clauses of the specification. Furthermore documentary evidence must be available to post-registration auditors to prove that such procedures are being followed. The discipline of the standard has had some effect on the contents of product specifica-tions and emphasised the distinction between **specified** properties

(guaranteed) and **typical** properties (published for the guidance of customers). It is generally accepted that the allocation of testing resources to
supporting lengthy and all embracing specifications is uneconomic for
suppliers (particularly of commodity products) and often of no value to
customers. However in individual cases the standard specifications published by the supplier may be considered insufficient to ensure consistency of performance in the customer's process and end-product. In
such cases it will be necessary for both parties to agree a special
customised specification. Possible consequences of such an agreement for
the supplier are additional testing and/or segregated storage. To avoid
any unnecessary effort and cost there is clearly a need for mutual
understanding between the parties on the significance of each clause
within a specification.

It is sometimes forgotten that in the field of plastics additives there is
often an intrinsic communication gap between producers and users. The
suppliers reside in the chemical industry where they operate chemical
processes and, as far as they are able manufacture products of consistent
chemical composition. On the other hand plasticiser users are most
commonly plastics processors whose quality managers and technologists
may not necessarily have any chemical background. They will be responsible for the details of quality management systems which ensure that
purchased additives show consistent performance in their formulated
products during compounding, fabrication and service. Since it is frequently not obvious whether the parameters included in suppliers'
specifications give any such guarantee there is always a need for a high
level of trust between the parties. Bilateral quality agreements sometimes
state explicitly that the supplier must inform the customer of any change
in raw material source (not generally applicable in the case of commodity
raw materials) or manufacturing process (even though such a change may
not have any effect on the specified parameters). It is usually the duty of
the supplier's technical service personnel to maintain the dialogue on such
matters.

The main part of this chapter covers in detail the various parameters
commonly included in plasticiser specifications, their inter-relationship
and their relevance to users' processes and end product performance.

8.3   CERTIFICATION

There is no inherent requirement in BS EN ISO 9000 registration for a
supplier to issue certificates with deliveries as proof that the product
conforms to the specification agreed with the customer. Indeed it is
reasonably argued that registration removes the need for these certificates

since the registered supplier's quality system must be of a sufficient standard to eliminate the possibility of a non-conforming material being delivered. In the early days of BS 5750 some suppliers regarded the elimination of the practice as one of the benefits offsetting the costs of registration. However even though this argument is generally accepted, for a variety of reasons there is an increasing trend to certify delivered products in many European countries.

One factor encouraging demand for certificates is competitive pressure between suppliers. A supplier whose material storage and data retrieval systems allow routine issue of certificates at no incremental cost is likely to agree readily to the demand without any discussion of its purpose and value. Once the practice is written into the customer's quality procedures it becomes very difficult for another supplier to challenge its basis. It may be the case that pressure for certification arises from large downstream customers such as car constructors who are one or more stages removed from raw material suppliers.

As already stated, most certificates obtained simply to prove that the supplier has delivered conforming product add nothing to the guarantee inherent in purchasing from a BS EN ISO 9000 registered source. However it may be that the data contained in a certificate can be of real value if used as input to the customer's process or to monitor the consistency of a supplier's product. In the latter case the customer's purpose could be equally served by requesting access to the supplier's statistical quality control (SQC) data if available. It is usually sufficient to confine such requirements to two or three key parameters selected from the full specification.

There is often considerable ambiguity over what is meant by a 'certificate of analysis'. The practicalities of storage and sampling may result in the certificate not representing with absolute precision the delivered product. In Germany there have been moves to recognise and clarify this by standardised classification of certificates as 'producer certificates' and 'producer inspection certificates'.[107] As usual satisfactory agreements are only achieved through close dialogue between supplier and customer on the scope and purpose of the information required, and the practical and economic constraints involved.

8.4   PLASTICISER SPECIFICATIONS

The parameters used to specify plasticisers can be considered in three categories.

Firstly there are physical properties which have easily measured characteristic values. These are useful as identifiers and conformance to specified values increases the probability that the material is what it purports to be. The most widely used such tests are refractive index, density and viscosity. Since different materials may have very similar values for these properties conformance does not give absolute proof of identity. Subjective assessment of odour and appearance against an expectation of what is typical are often used as supporting aids to identification.

Secondly there is a group of tests measuring properties which are affected by contaminants or the levels of characteristic impurities in the product. Such tests are relevant to the performance of the plasticiser in the customer's process and end product. They are performance orientated and do not seek to identify or quantify the impurities concerned. Examples are odour level, colour, electrical resistivity and windscreen fogging number.

Finally there is analysis of the chemical composition of the product by a variety of techniques. Specifications may contain clauses for minimum limits for product purity together with maximum limits for specific impurities believed to have a bearing on various aspects of performance.

The test methods cited in plasticiser suppliers' delivery specifications are generally national standards (BS, DIN, ASTM etc.) which may or may not have been harmonised internationally through ISO or CEN. In some cases suppliers will have found it necessary to introduce unilateral modifications of these methods or to use in-house methods to improve speed and accuracy. Consequent discrepancies between methods are a common subject of discussion between suppliers and users. Any significant differences in results obtained are generally resolved by controlled comparison.

Whilst there are in existence national standards specifying properties for some long established plasticisers (e.g. BS 4968, BS 5734) they are now rarely invoked. Some of the test methods which they cite have in industry been superseded by more modern techniques. In addition improved plasticiser manufacturing capability and more stringent market demands have caused many suppliers' specifications to outstrip these standards.

8.5 PLASTICISER SPECIFICATION PARAMETERS

Although this section reviews a wide range of possible parameters it is rare for plasticiser specifications to contain more than six or seven clauses except where there are special end use requirements. Some parameters are obviously only applicable to certain chemical types of plasticiser. The set most commonly used for phthalate esters is:

colour
refractive index

density
viscosity
assay or 'ester content'
acidity
water content

8.5.1 APPEARANCE

This is usually specified qualitatively as 'clear and free from matter in suspension'. Nonconformance could result from precipitation of an insoluble impurity from the product giving a cloudy appearance. Another potential cause of failure is external contamination by flakes of rust or surface coating material from tanks or pipes. When this occurs the amount of contaminant involved is usually minute. However because of the potential for causing visible defects in clear or pale coloured end products it is unacceptable in many applications. Hence it would be a cause of complaint whether or not explicitly covered in the specification.

One defect detectable by visual inspection of plasticisers is the presence of water. It may appear as cloudiness or fine droplets which coalesce on standing. When it occurs it is most often associated with incomplete drying of a tanker prior to loading. In the case of plasticisers with relative density <1 (including all the high volume commodity phthalates) it is usually confined to the lowest point of the tank and can be drained off through the foot valve. Since the solubility of water in most plasticisers is very low (typically <0.2%) and rates of solution and diffusion are correspondingly slow, the presence of localised water does not indicate that the specification limit for water content has been exceeded in the bulk of the material.

8.5.2 ODOUR

Whilst the human nose falls short of the olfactory performance of those of other species it is still an extremely sensitive instrument capable of detecting minute concentrations of airborne materials. The trained and experienced nose of a perfumer has quality control capabilities which cannot be matched by instrumental methods. Application of this level of ability is hardly required by plasticiser users. However a rapid assessment of odour coupled with appropriate experience can be very informative in many situations. Some purchasers of plasticisers specify that odour should be 'typical'. Since this is rather subjective, acceptance of such a clause may cause some concern to suppliers. However, provided the user's assessor is familiar with the normal range of variation of the product the requirement is valid. Deviation from a 'typical' odour could be indicative of the wrong material, an unusually high level of a characteristic impurity or the occurrence of external contamination. This would usually prompt analysis of the product, usually by gas chromatography.

For some applications such as food packaging it is important to control plasticiser odour in order to avoid tainting of the packaged contents. In such cases it may be necessary to assess the deviation of odour from that of an acceptable standard sample. This may involve the use of panels of selected assessors and analysis of their results by statistical techniques. The most rigorous procedures assess conferment of taint rather than the plasticiser odour. This requires compounding of the plasticiser sample followed by a period of contact with the foodstuff or a suitable simulant. Whilst elevated temperature may be employed to accelerate tainting, such tests are clearly too lengthy for acceptance testing.

The odour of any material, natural or synthetic, may be due to a combination of many components some of which may be present in minute traces. It is accepted wisdom in the fragrance industry that chemical analysis can never fully define or control odour. However where the major contributors to odour and their origins are known analysis (usually gas chromatography) can be used by plasticiser manufacturers to control product odour. This allows the possibility of including maximum levels in customised specifications.

Most plasticisers have very low characteristic odours which may be undetectable at ambient temperature but which become apparent in processing. For carboxylic esters the most common odour is that of residual alcohol feedstock not completely removed during the purification stages of manufacturing. In the case of DOP, for example this is 2-ethylhexanol. Plasticiser alcohols have characteristic sharp odours which are evident at ambient temperature if the concentration in the plasticiser is more than about 0.1%. Typical levels found in commercial products are much lower than thirty years ago and problems relating to alcohol odour are rare. When they do occur they are likely to be as a result of ester hydrolysis during or after compounding (See Chapter 3, 3.11) rather than being related to plasticiser quality.

It is known that other odours can occur in plasticised PVC as a result of complex interactions between the various ingredients of the formulation.[108] However this subject is outside the scope of this chapter.

8.5.3 COLOUR

The majority of plasticisers have a level of colour which falls in the range of water white to pale yellow. Plasticiser colour is most commonly measured on the platinum–cobalt scale (also known as the Hazen scale and the APHA scale) as described in ASTM D1209, DIN/ISO 6271 and BS 5339. These methods are designed for measuring very low intensities of yellow colour and produce a range of values up to 200 above which they are inaccurate. For more intense colours various alternative scales are used such as Iodine Colour Number (DIN 6162) and Gardner Colour. The latter takes into account the blue and red components of colour in addition to yellow.

Access to pure raw materials and improvements in manufacturing technology have enabled plasticiser suppliers to include very low values for colour in their delivery specifications. The difference between a Hazen colour of 15 and 50 is of real technical significance in few if any flexible PVC applications. Colour specifications are often driven more by manufacturing capability than by market need. Suppliers may use colour specifications to gain competitive advantage or to differentiate between standard and special grades of the same material. Once a user has adopted a low colour level into a purchasing specification relaxation of the limit to accommodate alternative suppliers is unlikely.

The impurities most commonly responsible for colour in plasticisers are probably organic chromophores. They are present at such low levels that any attempt at precise identification is unlikely to be successful. Colour is often associated with oxidation and exclusion of oxygen from the manufacturing process is considered a prerequisite for a low coloured product.

Whilst there is some tendency to regard low colour as a general indicator of high quality it is really quite unrelated to other quality parameters.

8.5.4 REFRACTIVE INDEX n_D^{20}

Refractive index is dependent on the wavelength of the refracted light as well as temperature and standard measurements are always carried out on a sample at 20°C using a sodium light source (sodium D wavelength). The method is standardised in DIN 51423 and ASTM 1045. Measurement of refractive index against a specified range is popular as a raw material check because of its speed and simplicity. However it is unrelated to performance parameters and even as a product identifier it is of restricted value.

A study of different suppliers' specifications indicates that DOP has a narrower range of refractive index (1.486–1.487) than almost any other plasticiser. Since DOP is one of the purest of industrial chemicals (typically containing more than 99.5% of a single component) this is not surprising. It is reasonable to assume that observed variation in the refractive index of the material is due mainly to the variability of the test. There are two reasons for the fact that other plasticisers have wider refractive index specifications than DOP. In the case of multi component products normal variation in the component proportions can affect physical properties (without necessarily having a significant effect on performance). The other reason for wide specifications for refractive index or any other parameter is lack of sufficient data for the supplier to have statistically established confidence in tighter limits.

Where a supplier has a range of chemically related products with overlapping refractive index specifications the significance of a correct value for a delivered product is clearly limited.

8.5.5 DENSITY

The density of liquids can be measured using simple equipment. Consequently density has often been used by plasticiser purchasers as a check on delivered material. It is most commonly specified at a temperature of 20°C in units of $g\,ml^{-1}$ ($kg\,l^{-1}$). Occasional confusion arises through the use of **relative density** (specific gravity) in specifications. These values need to be multiplied by the density of water ($0.998\,g\,ml^{-1}$ at 20°C) for comparison with density figures.

Various standardised methods are quoted in suppliers' literature including DIN 51757, ASTM D1045 and ASTM D941. The density range typically specified for DOP (0.983–0.985 g/ml at 20°C) is like refractive index variation more a reflection of test accuracy than indicative of any real variation in this essentially pure chemical. Most other plasticisers have somewhat wider specified ranges.

There is some overlap of density specifications for closely related plasticisers (although to a lesser extent than with refractive index). Hence a correct value for density does not constitute proof of product identity.

8.5.6 VISCOSITY

Where plasticiser viscosity limits are specified it is usually **absolute viscosity** figures which are quoted. Absolute viscosity (dynamic viscosity) is expressed in units of milliPascal seconds, mPa.s (see Chapter 2, 2.5.1). Less commonly it is the **kinematic viscosity** which is used, this being expressed in centiStokes. The relationship between the two quantities is:

$$CentiPoise = CentiStokes \times Density$$

Standard methods quoted in supplier's delivery specifications include ASTM D445, DIN 51 562 and DIN 53 051. The most common test temperature is 20°C although higher temperatures are often used with particularly viscous products.

For plasticisers other than those with a predominantly linear aliphatic structure (e.g. adipates) viscosity varies steeply with temperature under typical delivery conditions (see Table 8.1) and significant test error can occur in the absence of precise temperature control. Consequently the use of viscosity measurement on receipt as a rapid check against specification gives a risk of misleading results.

TABLE 8.1
VARIATION OF PLASTICISER VISCOSITY WITH TEMPERATURE AT 20°C $\pm$ 2°C

PLASTICISER	Viscosity (mPa.s)	
	at 18°C	at 22°C
Di-2-ethylhexyl adipate	12.3	14
Di-isobutyl phthalate	34.5	45.5
Di-(linear $C_9/C_{10}/C_{11}$) phthalate	51.5	63
Di-2-ethylhexyl phthalate	70	88
Tri-(linear C_8/C_{10}) trimellitate	112	138
Di-isodecyl phthalate	114	148
Tri-2-ethylhexyl trimellitate	270	340

Unlike refractive index and density, variability of viscosity for some types of plasticiser can be technically significant in relation to both the manufacturer's and the user's processes. With chlorinated paraffins viscosity increases logarithmically with chlorine content and viscosity specifications are broader than for other types of plasticiser (see Chapter 5, 5.3). The more highly chlorinated products can be difficult to transfer and process particularly when cold and here a maximum viscosity limit can be an important requirement for the user. The same is true of polyester plasticisers which do not have a precisely defined chemical composition (see Chapter 5, 5.6). Polyesters are commonly classified according to their viscosity which is related to their degree of polycondensation/molecular weight and to the concentration of unesterified hydroxyl end groups. Hence viscosity is a key clause of the specification for these products.

At the other extreme are single component plasticisers such as DEP, DOP and DOA. As with refractive index and density any observed variation in viscosity with such products are more likely to be due to test error than to product variability.

A large proportion of plasticisers lie between these extremes in being multi-component as a result of being produced from mixed feedstocks. Examples are DIDP, most linear phthalates and phosphates of iso-propylated phenol. Consistency of plasticiser viscosity can be taken as an indication of consistency of feedstock composition. In practice variation of composition with this type of product is technically insignificant.

8.5.7 ELECTRICAL RESISTIVITY

Manufacturers of insulating PVC compounds for electrical cables need to eliminate the possibility of variability in any of the compound ingredients having an adverse effect on the electrical performance of their products. Most users have found by long experience that plasticisers are not a significant source of variation. Consequently it is unusual for their purchasing specifications for plasticisers to include clauses explicitly controlling electrical performance. However with the introduction of quality management systems some plasticiser purchasers consider that the specification of a minimum value for volume resistivity (V.R.) is appro-priate. In view of the evidence that PVC compound V.R. is only weakly dependent on plasticiser V.R. (see Chapter 6, 6.2.1), it is possible for such a specification to be set quite loosely without risk to the performance of the compound.

One standard test method which has been invoked in plasticiser purchasing specifications is ASTM D1169. The scope of this standard is measurement of the specific resistance of insulating liquids rather than plasticisers for insulating compounds. For the most accurate measure-ments the standard stipulates the use of a guarded electrode cell although a simpler 2-electrode cell may be used for routine testing. These measure-ments are extremely susceptible to error as a result of sample contamina-tion and great care is required in cleaning the test cells between tests and in transfer of samples. The ERA cell[109] is a simple compact apparatus incorporating two electrodes in a single assembly. This not only facilitates cleaning but allows immersion of the electrodes directly into liquids held in sample jars thereby reducing the risk of contamination during transfer. Whilst such techniques are not standardised they offer a practical option for screening materials against serious defects.

8.5.8 THERMAL STABILITY

It is unusual for thermal stability to be included in plasticiser specifica-tions since for most types this is not a relevant issue. Chlorinated paraffins are an exception and their stability is expressed in terms of their pro-pensity for generating hydrogen chloride – for example the % mass of HCl split of after 4 hours at 175°C.[110]

8.5.9 VOLATILITY

Some form of volatility test may be specified as a means of ensuring that the plasticiser is free from volatile components. Such specifications are unusual and where the identity of volatile impurities is known it is generally more satisfactory to specify maximum limits based on gas chromatographic analysis.

8.5.10 WINDSCREEN FOGGING

Purchasers of plasticisers for use in automotive interior trim components often specify a maximum limit for the level of windscreen fogging produced by the uncompounded plasticiser. The fogging test method and its significance are discussed in Chapter 3, 3.3.4.

Variability in fogging performance of a plasticiser is due to the presence of volatile impurities such as unesterified alcohols. Since such impurities are quantifiable by gas chromatography, plasticiser suppliers are likely to use this technique for quality control of 'non fogging' grades of plasticiser rather than rely on fogging tests which are time consuming and have poor reproducibility.

The remaining specification parameters covered here are all based on chemical analysis of the product.

8.5.11 WATER CONTENT

Nearly all plasticisers are liquids which are hydrophobic to varying degrees. Approximate solubility limits of water in a variety of different plasticisers are shown in Table 8.2.

TABLE 8.2
APPROXIMATE SOLUBILITY OF WATER IN PLASTICISERS AT 20°C

PLASTICISER	Solubility (%mass)
tri-2-ethylhexyl trimellitate	0.1
di-2-ethylhexyl phthalate	0.15
di-2-ethylhexyl adipate	0.2
di-isobutyl phthalate	0.4
diethyl phthalate	0.7
triethyleneglycol di(caprate/caproate)	1.0

Maximum water content is included in suppliers' specifications for most plasticisers a typical limit being 0.1% by weight. For some plasticisers this is only just inside the solubility limit and levels in excess of specification would be expected to produce visible cloudiness or droplets.

Various standard test methods are used by plasticiser suppliers including BS 2511, DIN 51 777, Part 1 and ASTM E203.

Water contents specified by plasticiser suppliers are related at least as much to manufacturing capability as to users' technical requirements. PVC resins are larger potential contributors to the water content of a plasticised composition than plasticisers. Specifications for suspension resins range from 0.2% to 0.4% maximum water level compared with 0.05% to 0.1% for phthalate plasticisers. Within these limits it appears that water has no effect on the processing or properties of the compounded products. It is assumed that rapid and essentially complete volatilisation occurs during the high temperature stages of processing.

The use of plasticisers in moisture-cured polyurethane sealants provides an unusual example of a real need for very low water content to be specified.

8.5.12 ACIDITY/ACID VALUE

Ester plasticisers always contain traces of the acids from which they are derived (usually in the form of partially esterified acids). Imposition of a maximum acidity specification ensures that the esterification reaction has been completed and that the product has been purified of acidic impurities which might have adverse effects on performance.

Acidity is measured by titrimetric methods as described in BS 4835, ASTM D 1045 and DIN 53 402. Plasticiser suppliers specifications usually refer to **acid number**. This is the number of milligrams of potassium hydroxide required to neutralise 1 gram of the plasticiser ($mg\,KOH\,g^{-1}$). Values typically specified range from 0.05 for phthalates to 2.5 for polyesters which typically contain a proportion of unreacted acid end groups. An alternative way of expressing acidity which is sometimes encountered is in units of $meq\,kg^{-1}$ (the number of milliequivalents of acid contained in 1 kg of plasticiser).

Yet another expression of acidity is as the percentage of a specific acid (e.g. phthalic) equivalent to the observed acidity.

A typical phthalate specification limit of $0.07\,mg\,KOH\,g^{-1}$ is the same as $1.25\,meq\,kg^{-1}$ or 0.01% phthalic acid. In reality the acidity would be nearly all present in the form of a monoalkyl phthalate for which the specification would represent a maximum limit of 0.035%. This concentration of a weak organic acid is far too low to cause any adverse effect in normal technical applications. However suppliers sometimes use different acidity specifications to differentiate between 'standard' and 'special' grades of the same product. An example of the latter is DOP for use in medical PVC compounds.

8.5.13 Hydroxyl value

With the adoption by plasticiser manufacturers of gas chromatography for quality control this non-specific wet chemical method for measuring -OH group content has largely fallen into disuse. However it is still relevant to the characterisation of non-stopped hydroxy terminated polyester plasticisers.

8.5.14 Saponification number

In the far off days before gas chromatography the most convenient way of checking whether an ester plasticiser had the correct composition was to measure its saponification number. Acceptable upper and lower limits were included in the specifications for these products.

Saponification number is defined as the number of milligrams of KOH consumed in saponifying 1 gram of the ester. Its value is proportional to the number of ester groups in the molecule and inversely proportional to molecular weight.

A parameter derived from the saponification number was the **ester content**. This is the observed saponification number expressed as a percentage of the theoretical value for the pure material. Contamination by a material containing a higher proportion of ester groups, for example DBP in DOP would give as ester content in excess of 100%. Therefore it was necessary to specify an upper as well as a lower limit, a typical range being 99–101%.

A correct 'ester content' derived from saponification number measurement was no guarantee of correct composition since the same saponification number would be shared for example by DOP, DIOP or any phthalate mixture having the same average molecular weight. The replacement of saponification number measurement by gas chromatographic analysis has removed these limitations. The consequences of this are discussed in the following section.

8.5.15 Assay/purity

For all but a few plasticisers with molecular weight below about 1000 gas chromatographic (GC) analysis can measure precisely the percentages of specific single components or selected groups of components. Hence if the characteristic chromatogram peak or group of peaks comprising the 'pure' product are defined, all others are considered as impurities.

Determination of relative peak areas then leads to a figure for percentage purity (assay). For absolute accuracy the difference in response of the detector system to different components should be taken into account although some routine methods avoid this.

Despite the widespread inclusion of GC-derived purity limits in plasticiser specifications there are no standard methods available and test details can vary considerably between suppliers. The following method published in BASF's plasticiser literature[111] indicates suitable conditions for determining the purity of DOP and similar plasticisers.

Separating column:	Silicone SE 52 (5%) on Chromosorb® WHP 100–120 mesh, stainless steel (e.g. V2A), 2 m long
Inner diameter:	2.17 mm
Outer diameter:	3.17 mm
Temperature:	Injection port 290°C
	Column oven 5 minutes at 230°C, then at a rate of $10°C\,min^{-1}$ to 270°C followed by 25 min isothermal
	Detector 290°C
Carrier gas:	Nitrogen, white spot, ca. $1.8\,lh^{-1}$*
Detector:	FID (fuel gases: hydrogen ca. $1.8\,lh^{-1}$* synthetic air ca. $18.0\,lh^{-1}$*
Evaluation:	Percentage area

* Guide values only (need to be optimised for instrument).

For a simple chemical like DOP the definition of purity by such a method is relatively unambiguous. However the majority of plasticisers are mixtures of closely related components often in very large numbers. Here it is helpful to have a knowledge of production processes and possible variations in raw materials in deciding which peaks represent the 'pure' material. Some suppliers prefer to use the old term 'ester content' in preference to 'purity' or 'assay' in order to allow some reasonable flexibility in interpretation. An increasing number of plasticiser users now have access to GC facilities. Because of the diversity of possible test conditions and particularly because of differences in interpretation of results there is potential for misunderstandings between customer and supplier unless open and clear communication on the subject has been established.

Variations in plasticiser composition revealed by GC are often of no relevance to technical performance. For example, contamination of DOP by 5% of another general purpose plasticiser, DINP or DIDP, would have no detectable effect on either its processing properties or the properties of plasticised PVC end products. However GC would show it to have failed a typical supplier's specification of 99.5% minimum assay. In order to deliver the material the ISO 9000 registered supplier would need to request a waiver of specification by the customer. If, as is now likely the customer were also an ISO 9000 company a set procedure would have to be followed before a non-conforming material could be accepted. If this proved in any way irksome the customer might prefer simply to transfer the order to an alternative supplier. All along the customer's technologists might have been quite confident that the rejected material would have been technically acceptable.

8.5.16   SPECIFIC IMPURITY LEVELS BY GC

If the conditions are adjusted to give good separation of components appearing before the main peaks GC can be used to quantify specified volatile impurities. Hence it can be useful for quality control where any such impurity has a **known** effect on some aspect of product performance. For example residual alcohol in a phthalate plasticiser is the clearest potential source of odour and taint. Alcohol level also needs to be controlled even if not explicitly specified, for 'non-fogging' grades of plasticiser.

Other specific impurities determined by GC may be included in bilateral specification agreements between suppliers and individual customers but are seldom if ever included in published specifications. The method used in any such case is likely to have been developed by the plasticiser supplier. Whilst GC can be a cost-effective technique with a number of useful applications for the plasticiser user, its capability of revealing minor components at very low levels can sometimes create unnecessary concerns. Again these can only be avoided through good technical communication between supplier and customer.

8.5.17   OTHER IMPURITIES

Plasticiser suppliers may asked by some customers to state that their materials are free of heavy metals. A more precise requirement would be conformance to a specification for maximum tolerable levels of specific heavy metals. Such requirements relate to application areas such as medical products, food packaging and toys.

It may be sufficient for the supplier to base assurances on the fact that the raw materials and manufacturing processes introduce no possible source of heavy metals. However if a formal quantified specification is agreed then it will be necessary to carry out testing at sufficiently regular intervals to satisfy accredited auditors if a ISO 9000 registered system is in force. Methods of determining metal contents at levels around 1 p.p.m. ($mg\,kg^{-1}$) (e.g.) are usually based on atomic absorption spectroscopy.

8.5.18 ANTIOXIDANT LEVEL

For use in PVC cable compounds operating at elevated temperatures plasticisers are usually supplied containing dissolved antioxidant, typically Bisphenol A (see Chapter 2, 2.3.8). Since omission of the antioxidant could drastically impair the thermal stability of the compound, users may require a formal specification for antioxidant level. The nominal level of Bisphenol A in high temperature plasticisers is, by common practice 0.5% by weight although this is not critical. Specifications may be set as either a range (e.g. 0.4 to 0.6%) or as a minimum level (e.g. 0.4% minimum). Test methods will be suppliers methods based on e.g. liquid chromatography (HPLC) or UV absorbance of an aqueous sodium hydroxide extract from the plasticiser.

8.6 ANALYSIS OF PLASTICISERS

8.6.1 IDENTIFICATION OF AN UNKNOWN PLASTICISER

The use of gas chromatography for determining the purity of a plasticiser or for measuring levels of specific impurities has been discussed in 8.5.14 and 8.5.15. GC is also the most useful single method of confirming or determining the identity of the great majority of plasticisers. Polyesters and chlorparaffins are two notable groups not amenable to separation by GC. For a given type of chromatography column and set of conditions each plasticiser has a characteristic fingerprint of peak positions and peak area distribution by which it can be identified against standards.

Figure 8.1(a) to 8.1(g) compare typical gas chromatograms of a number of ester plasticisers, both single component materials and materials derived from mixed feedstocks. They were obtained using a 50 m wide bore (0.53 mm i.d.) capillary column under the following conditions:

Temperature: Injection port 250°C.
                 Column oven 80°C increasing by $8°C\,min^{-1}$ to 320°C
                 followed by 15 min. isothermal.
                 Detector 300°C
Carrier gas: Helium
Detector: FID

Single component products	*Description**	*Mixtures*	*Description**
		isodecyl diphenyl phosphate	5.7.3
DOA	5.4	DINP	4.6.2
DOP	4.6.1	DIDA	5.4
DOS	5.4	DIDP	4.6.2
TOTM	5.8.2	DITDP	4.6.2
		T810TM	5.8.4

* Chapter sections describing product compositions.

Although GC has been used by some PVC compounders and processors to check the identity of plasticiser deliveries it is generally regarded as being too sophisticated and time consuming to be used routinely for this purpose. The need for such checks is nowadays reduced by the reliability of suppliers' ISO 9000 accredited quality management systems.

Since it is possible for quite different materials to have very similar retention times the use of GC on its own does not eliminate uncertainty of identity. This limitation can be overcome by coupling it with a second technique to identify the separated components. The combination of GC with mass spectroscopy is generally the most informative such method.

Liquid chromatography (HPLC) is one of the most widely used tech-

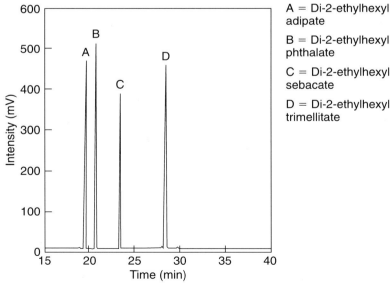

A = Di-2-ethylhexyl adipate

B = Di-2-ethylhexyl phthalate

C = Di-2-ethylhexyl sebacate

D = Di-2-ethylhexyl trimellitate

Fig. 8.1a Gas chromatograms of single component plasticisers.

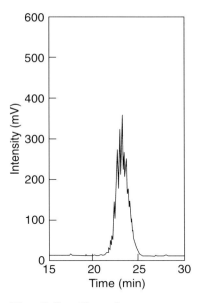

Fig. 8.1b Gas chromatogram of di-isononyl phthalate (polygas-derived).

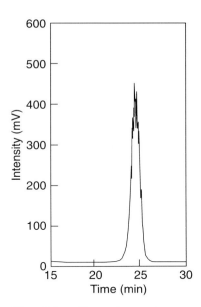

Fig. 8.1c Gas chromatogram of di-isodecyl phthalate.

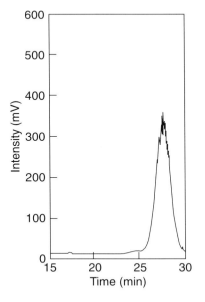

Fig. 8.1d Gas chromatogram of di-isotridecyl phthalate.

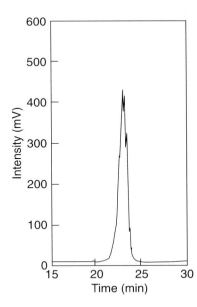

Fig. 8.1e Gas chromatogram of di-isodecyl adipate.

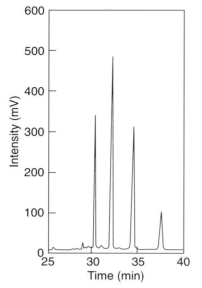 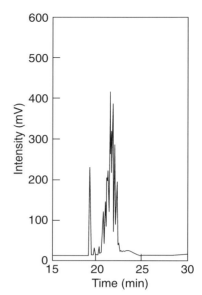

Fig. 8.1f Gas chromatogram of tri (n-C_8/C_{10}) Trimellitate.

Figure 8.1g Gas chromatogram of isodecyl diphenyl phosphate.

niques for separating and identifying plastics additives.[112] It has the capability of separating with high resolution high molecular weight materials which are not amenable to GC analysis, including polyester plasticisers. Like GC it can be coupled with mass spectrometry to allow precise identification of unknown components. However except in special cases it is more convenient to use GC rather than HPLC for plasticiser analysis.

So far no mention has been made of infrared spectroscopy as a technique for analysing plasticisers. IR spectroscopy works by absorption of radiation with frequencies characteristic of specific groups within a molecule. Hence although it is capable of showing, for example the presence of a phthalate ester it does not distinguish clearly between the many possible dialkyl phthalates. Therefore it provides only a limited identification of an unknown sample. IR can be useful for classifying a material sample prior to more detailed analysis by GC. The particular merit of IR is that unlike chromatographic techniques it can (partially) identify and in some cases quantify additives in solid plastic compositions without the need for the extraction or dissolution steps discussed in the next section.

8.6.2 ANALYSIS OF PLASTICISED COMPOUNDS

The most widely used techniques for qualitative or quantitative analysis of plasticisers in plasticised compounds require separation of the plasticiser from the polymer. This is achieved either by solvent extraction from a finely divided sample of compound or by complete dissolution of the compound in a suitable solvent followed by reprecipitation of the polymer.

Most plasticisers can be removed almost quantitatively from PVC by Soxhlet extraction or direct refluxing with diethyl ether for several hours. Extraction of the most permanent polyester plasticisers may require extended reflux times or alternative solvents. With the solution/reprecipitation method PVC compounds can be dissolved in THF (the toxicity of this solvent necessitates the use of exposure control measures). The PVC polymer can then be reprecipitated by the addition of n-hexane. The ether or THF solution is injected onto the GC column following adjustment of concentration if required.

The time and labour involved in separating plasticiser from the polymer can be a disincentive to analysis in situations where it could contribute valuable information. Surprisingly reliable results have been obtained over a long period by using a short cut in which PVC compound (0.3 g) is first dissolved in THF (10 g), insoluble material is allowed to settle and then the THF solution *including PVC* is injected onto the GC column. By calibration against standard compositions the method is capable of giving quantitative results. The method causes some fouling of the column packing by PVC residues. This can be dealt with by regular replacement of the first few centimetres of the packing.[113]

An elaboration on this method which has been used to overcome fouling is cold injection of the THF solution into a small Curie point pyrolysis coil. The coil holding a drop of the solution is then placed in the injection port of the GC and raised instantaneously to a temperature of 368°C[113] as power is applied. This injection technique requires special adaptation of standard GC equipment.

A relatively new approach to the problem of long extraction times for plasticisers and other additives is the use of supercritical fluid extraction coupled with supercritical fluid chromatography. Here the additives are extracted from a finely divided sample using liquid carbon dioxide which because of its high diffusivity gives far faster extraction than conventional solvents. Following release of pressure and evaporation the CO_2 functions as the GC carrier gas. Hunt *et al.*[114] were able to obtain reproducible quantitative results for the di-iso-octyl phthalate context of PVC using extraction times of less than 20 minutes.

8.6.3 DETERMINATION OF PLASTICISER CONTENT AS A QUALITY CONTROL FOR PVC COMPOUNDS

In the manufacture of plasticised compounds it is essential to ensure that any variation of plasticiser content falls within acceptable tolerances. A rapid and reliable method for in-plant testing is desirable. Unfortunately the methods discussed in the previous section are too slow and too elaborate for this purpose. Measurement of some physical property dependent plasticiser content may be helpful although this approach has serious disadvantages. For example determination of softness number requires the moulding and conditioning of test specimens. If, as likely here, conditioning times are very short there is a risk of seriously misleading results (see Chapter 2, 2.6.1.5)

A technique which is now well established for quality control in the production of flexible PVC compound is low resolution broad line nuclear magnetic resonance (NMR) spectroscopy. This is quite distinct in its scope and simplicity from high resolution NMR which is a powerful technique for detailed determination of molecular structure. Although low resolution NMR has been used successfully for a number of years by some large PVC compounders and processors it is still far from universally known in the industry. It works by producing a signal which is proportional in

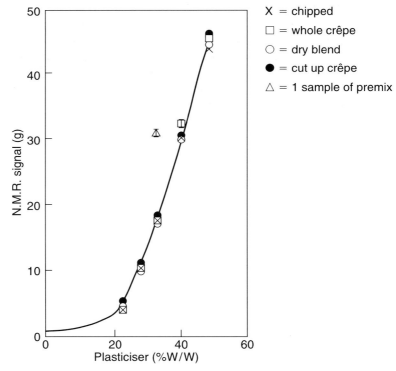

Fig. 8.2 DIOP plasticiser in PVC resin. Various concentrations of DIOP all milled for 10 min. 95% confidence limits of two of the points are illustrated. The possible errors on all NMR measurements are of the same order.

strength to the number of hydrogen atoms contained in the *mobile* molecules present in a material (plasticisers etc.). The response from hydrogen atoms in the immobile polymer molecules is screened out. The technique was adopted for use with plasticised PVC in the nineteen seventies after having become established in the edible oil industry as a routine method for measuring the oil content of seeds. Commercial quality control instruments are robust and simple to use. A weighed sample in any convenient form (e.g. compound granules) is conditioned at a standard temperature for a short time after which an instant digital reading related to the plasticiser content is obtained. A typical time for the whole process is 5 minutes.

The observed signal per gram depends on the hydrogen content of plasticisers and any other mobile additive molecules present. It is reduced by strong immobilising interaction between the polymer and plasticiser and therefore is influenced by the degree of fusion which has been achieved. A PVC–plasticiser cold premix gives a much stronger signal than a dryblend in which PVC–plasticiser association has occurred at a molecular level. Melt processing of the dryblend then gives no further reduction of the signal. Because of the variables involved the NMR method requires calibration using control compounds of appropriate composition. It then becomes a rapid, reliable and practical indicator of consistency. Figure 8.2 shows a calibration plot for PVC containing different levels of di-iso-octyl phthalate. Results for a variety of sample forms are included – dryblends, milled hide, chips and one premix.[115]

CHAPTER 9

Health, Safety and the Environment (HSE)

Note: Throughout this chapter the abbreviations DEHP and DEHA are used for di-2-ethylhexyl phthalate and di-2-ethylhexyl adipate rather than their common industrial designations DOP and DOA.

9.1 INTRODUCTION

So far in this book we have considered the technical and economic factors influencing the selection of plasticisers for particular applications. The very wide range of technical possibilities open to a formulator is limited by cost considerations to a small list of practical options. The great majority of plasticiser consumption is limited to the three 'general purpose' phthalates which are used in flexible PVC for most of its applications.

The very maturity of the technology of external plasticisation and of the plasticisers themselves means that most users require less technical advice from suppliers than was the case thirty years ago. As discussed in the last chapter, consistency of quality of these familiar products has now assumed far greater importance for many customers.

If the only concerns of plasticiser users were to procure consistent technically acceptable materials at prices which allowed their end products to be competitive then future patterns of use would probably be very stable. However, nowadays the issues which dominate the agenda in any discussion of the future of plasticisers (and many other industrial chemicals) are those relating to human health and safety and to environmental impact. Public awareness and concern over such issues is now far greater than in the rapid growth era of plasticisers. Legislation reflecting this concern can place specific or broad restrictions on the use of chemicals. In addition sustained adverse publicity, whether or not scientifically substantiated, can influence a converter's selection of raw materials.

Ideally approvals and restrictions should be based solely on scientific assessment of the risks involved in using the different materials competing for a particular application. However in the real world politics in various forms enters the picture, sometimes with unintended results.

During the nineteen eighties plasticisers were exposed to unprecedented scrutiny as a result of research in the USA showing the potential of DEHP and DEHA for causing cancer in rodents. In response to this large programmes of research were initiated to investigate the possible carcinogenic risk of plasticisers for humans. The eventual conclusions were sufficiently clear to satisfy legislators in Europe that the materials concerned posed no such risk. Before this stage was reached there was a period of concern and uncertainty which had effects on the plasticiser market. A fragmented response by plasticiser users and **their** customers caused some disruption in the pattern of plasticiser usage. In Europe these concerns have largely faded and the agenda has moved on to other potential health risks and to the environmental impact of plasticisers.

All industrial chemicals are now subject to harmonised European Community legislation which requires standardised classification of their hazards and communication of this information to users. Whilst this will eventually create a framework against which the risks involved in using any plasticiser can be assessed the situation is currently unbalanced since most hazard testing has been confined to a handful of key materials. This is an important factor in understanding why attention is always focused on the same small selection of plasticisers. Because of their industrial dominance the phthalates have been studied in far greater detail than other types. Of this group DEHP has received the most attention partly because it has long been the major product. It also happens to offer the researcher a relatively simple subject for study since unlike most plasticisers it contains a single chemical component. Whilst this focus is understandable and logical it has sometimes had the effect of suggesting that other types of plasticiser are 'safer' alternatives whereas in reality their toxic hazards are relatively unknown. In addition the absence of such materials from reports of environmental contamination is simply a reflection of their relatively low consumption.

Large scale demand for plasticisers depends above all on the continuation of flexible PVC as a major thermoplastic material. The outcome of the ongoing debate over the environmental acceptability of PVC will override all other influences on the long term future of plasticisers. Hence any balanced treatments of the HSE aspects of plasticisers must include a discussion of the arguments surrounding PVC.

This final chapter reviews these influences and place them in perspective. In contrast to earlier chapters it covers a very dynamic area where approvals and restrictions of individual materials could change very quickly. Hence it makes no attempt to give definitive statements on the legislative status of different plasticisers since this could quickly become dated and misleading. It is intended rather to provide a helpful background for understanding future developments.

9.2   THE NINETEEN EIGHTIES – THE CARCINOGENICITY QUESTION

Prior to the nineteen eighties the substances in common use as plasticisers had been well studied with regard to both acute and chronic toxicity. Acute toxicity indicates the effect that a single dose of the substance will have on a specified living organism. It is expressed as the LD_{50}, that is the dose which is lethal for 50% of test animals referred to their body weight. A typical plasticiser such as DEHP has an extremely low acute toxicity with LD_{50} in excess of 30,000 mg/kg. Putting this into perspective, ethanol has acute toxicity an order of magnitude higher (LD_{50} = 3,300 mg/kg).

Chronic toxicity is an indication of the effect on an organism of exposure to the substance over a long period. It is conventionally assessed by daily administration to test animals over a period of up to two years. For reasons discussed earlier the majority of such studies on plasticisers have involved DEHP. Tests carried out before 1980 showed no evidence of chronic toxic effects. In addition laboratory results and a wealth of experience of safe use in practice showed an absence of irritant effects on the skin and mucous membranes. Taken together the available evidence showed no cause for concern and the subject of plasticiser toxicity had a low profile.

In 1980 this picture changed with the publication in the USA of results of a two year feeding study under the National Toxicology Program (NTP)/National Cancer Institute (NCI) Bioassay Program. This study involved feeding rats and mice with diets containing extremely high levels of DEHP and DEHA. The dose levels (corresponding to a human intake of 1/4-litre per day and 1/2-litre per day respectively) were only made possible by the extremely low acute toxicity of the two materials. DEHP was an obvious subject for such a study on account of its industrial dominance and chemical simplicity. Closely related dialkyl phthalates were not included. DEHA, also a single chemical entity, was of particular interest because of its widespread use in food packaging film.

These findings brought considerable attention to bear on the two specific plasticisers concerned. The International Agency for Research into Cancer (IARC), an agency of the World Health Organisation used these results as a basis for classifying DEHP as 'probably carcinogenic to humans'. DEHA was placed in the class 'not classifiable as to its carcinogenicity to humans'.[116] In the wake of this, moves against DEHP were made by some legislators and unilaterally by some downstream consuming companies anxious about the implications of using a material apparently under suspicion. In the case of DEHA concern was specifically confined to its use in food wrapping film. On a rational level this was taken into account in the deliberations of the EC Scientific Committee on Food (SCF – see 9.5.1). It also gave rise to a public clingfilm scare which was fuelled both by uninformed media attention and by sectional interests within the packaging industry.

The other response to the NTP/NCI findings was an outburst of industrially sponsored research to determine their significance to human health. It was argued that since DEHP had been used on a large scale worldwide for more than thirty years without any indication of a link to human cancer the relevance of the results to human risk was questionable. In Europe the main work on this subject was co-ordinated by a group belonging to CEFIC, the European Chemical Industries Association. This group subsequently adopted the title ECPI, European Council for Plasticisers and Intermediates. This detailed research led to the following conclusions:

1. DEHP and DEHA are not genotoxic, that is to say unlike direct carcinogens they do not react with genetic material.
2. The mechanism by which they cause tumours after repeated high dosing of rodents is believed to be peroxisome proliferation, that is a large increase in the sub-cellular microbodies causing oxidative degradation of metabolites. It has been pointed out that common hypolipodaemic drugs which have been used safely for many years for long term treatment of humans are known to show this effect in rodents.
3. The results are specific to rodents. Feeding of plasticisers to species metabolically closer to humans does not cause peroxisome proliferation or liver tumours.
4. There is no significant difference in effect between DEHP and the alternative general purpose phthalate plasticisers.

The evidence from these detailed studies was taken into account by assemblies of scientific experts having the responsibility of advising legislators. In 1986 the German Advisory Body on Environmentally Relevant Existing Substances (BUA) concluded that the NTP studies had no relevance to man and that there was no basis for suspecting chronic damage by exposure to DEHP at environmentally relevant concentrations.[117] Essentially the same conclusion was reached in 1990 by an expert group appointed by the European Commission. In this case consideration of DEHP was prompted by action taken in Denmark under the provisions of the Dangerous Substances Directive (this is discussed in some detail in the next section). Denmark had in 1987 introduced national legislation requiring containers of DEHP to be labelled as carcinogenic and irritant. Having taken this action Denmark was then required to submit supporting evidence to the European Commission for a ruling. The Commission was in turn obliged to reach a conclusion without delay and then to take appropriate action. Having considered all evidence, including the results of the research described above the Commission decided that the Danish labelling requirement was not justified and should therefore be withdrawn.[118]

It is probable the Danish labelling requirement was based squarely on the fact of the IARC classification of DEHP. In this case it appears that there was a misunderstanding of the purpose and limitations of such a classification.[119] The preamble to the IARC monographs makes it clear that their purpose is to provide a warning signal of a possible risk. This then identifies the material concerned as a priority subject for expert review, risk assessment and then, if appropriate regulatory action. It was such a review and risk assessment which enabled the European Commission to reach its decision.

Following these events concerns over plasticiser carcinogenicity had largely faded in Europe by the early nineteen nineties and the agenda had moved on to consideration of other types of hazards. For a variety of reasons the situation is rather different in the USA where an authoritative statement covering the question of DEHP cancer risk has not been available and there have been calls for further research. During the nineteen eighties concern over the safety of DEHP was a factor favouring substitution by alternative plasticisers (predominantly DINP and DIDP) on both sides of the Atlantic. In Europe the effect was temporary and in most cases technical and economic factors took over again as the factors influencing plasticiser selection. However there remain some examples of DEHP avoidance dating from the previous decade, either in the form of precautionary legislation which has never been rescinded (e.g. the Swiss ban on the use of DEHP in toys) or unilateral bans by individual processors anxious to avoid any adverse publicity associated with DEHP. Lack of evidence of reduced risk resulting from the substitution has not been a consideration in such cases.

In the USA there is legislation relating to DEHP based on its IARC classification but this is confined to the state of California. 'Proposition 65', approved in 1986 prohibits anyone from wilfully and intentionally exposing an individual to any material included in a list of 'known carcinogens' without first giving a clear and reasonable warning. DEHP was added to this list on 1 January 1988 as a consequence of having been included in an

IARC monograph as a possible carcinogen. Had this been the end of the matter the use of DEHP in products sold in California would have been severely curtailed since cancer warning labels would obviously have acted as consumer repellents. However Proposition 65 dispenses with the requirement for a warning for exposures that pose 'no significant risk assuming lifetime exposure at the level in question'. The Californian Health and Welfare Agency has set this risk threshold at a level of 80μg/day for DEHP (a decision which recognises the fact that DEHP is not genotoxic).

Measurement of consumer exposure for individual products containing DEHP would be a daunting task. However the Chemical Manufacturers Association in the USA has developed a method whereby producers of PVC wallcoverings, flooring etc. are able to make scientifically informed estimates of exposure levels.[120] Average daily absorption of DEHP by skin contact and indoor inhalation are calculated by reference to standard tabulated data. To date application of the CMA method to such products has given values well below the stipulated 80μg/day no risk level. Consequent conclusions on the lack of need for labelling have not been challenged.

9.3 THE DANGEROUS SUBSTANCES DIRECTIVE

As mentioned in the previous section the Danish legal requirement for DEHP to be labelled as a carcinogen and its overruling by the European Commission came about because of the **Dangerous Substances Directive** (DSD). In full this is entitled the 'Directive on the approximation of the laws, regulations and administrative provisions relating to the classification, packaging and labelling of dangerous substances'. The parent directive 67/548/EEC which was implemented on 1 January 1970 has subsequently been the subject of twenty adaptations and seven amendments. The Seventh Amendment (93/21/EEC) implemented on 31 October 1993 has been of particular consequence for plasticisers as will be discussed later.

The purpose of this legislation is to ensure that the users of a substance (a chemical element or compound or a defined multicomponent composition produced by a chemical process) are alerted to its hazardous potential so that appropriate precautions can be taken to allow its safe use. This is accomplished by ensuring that:

(a)   the hazards of a substance are properly defined
(b)   safety information is supplied with the substance to the user

(c) packaging is suitable and secure
(d) the packages carry distinctive cautionary labelling

The means of complying with each of these requirements are specified in detail. A similar set of requirements are imposed on the supply of **preparations** (physical mixtures of substances) by the **Dangerous Preparations Directive** (DPD). The parent directive was implemented on 7 June 1991. If a preparation contains substances with known hazards then (depending on their concentration and the level of hazard) it may be necessary to classify the hazards of the preparation. In the context of plasticisers 'preparations' include plastisols, emulsions, solid compound granules and powder compounds. Any hazards of solid preparations need to be included in safety data sheets but not in their container labels. Products fabricated from such preparations are outside the scope of the regulations.

As these directives and their various amendments have been implemented, conforming legislation has been progressively introduced in the member states of the European Union. In the UK a key piece of legislation is the Chemicals (Hazard and Packaging) Regulations enacted in 1993 and known as 'CHIP'. Since these omitted the requirements of the recent Seventh Amendment to the DSD they have had to be replaced by an amended set of regulations (CHIP 2) which are expected to come into force early in 1995.

The directives and conforming national legislation require that chemical substances and preparations should be classified according to their physico-chemical, toxicological and eco-toxicological (environmental) effects. Where appropriate standard 'risk phrases' (warning of the hazard) and 'safety phrases' (indicating precautions to be taken by the user) must be allocated.[121] The supplier is then obliged to include these in labelling containers of the substance or preparation and in the safety data sheets supplied with the material. Bulk storage and transport of the materials are not covered by this legislation.

The job of classifying 1500 hazardous substances has already been carried out by the EU. These are catalogued in the 'Approved Supply List' (ASL). For these substances it is mandatory to use the EU allocated classification. For other substances (and preparations) the supplier is responsible for classification following a review of all relevant available data. At present the ASL does not include any materials commonly used as plasticisers.

The Fifth Amendment of the DSD, implemented in 1981 introduced a pre-marketing notification of new substances. A 'new substance' is any material not included in the European Inventory of Existing Chemical

Substances (EINECS). This is a list of substances marketed in commercial quantities between the reference dates of 1 January 1971 and 1 January 1981. Incorporation of this requirement into national laws placed an immediate obligation on suppliers to generate prescribed data covering physico-chemical, toxicological and environmental effects before introducing any new product to the market even in small quantities. The amount of data required considerably exceeded the information for very many of the substances listed in EINECS which in the short term escaped the need for such testing. Consequently there was a sudden escalation in the cost of test marketing new products. Hitherto it would have been possible to produce 1 tonne of a new plasticiser and to supply drums to customers for trials having made estimates of hazards by analogy with the known characteristics of closely related materials. The cost of testing to comply with the DSD in such a case are currently in the region of £100,000. This has inevitably acted as a constraint on the commercial introduction of new plasticisers since 1981.

Existing chemicals (those listed in EINECS) are gradually being brought into the net of testing requirements but since the number involved is enormous the process is being carried out on a priority basis. Suppliers of large tonnage materials (>1,000 tonnes per annum per supplier) have submitted standardised data (the so-called 'HED set'). HED sets for approximately 1,400 such substances are filed and of these 50 have been selected by the EU for risk assessment. Dibutyl phthalate has been included in this group because of evidence of its aquatic toxicity. Once risk assessments are complete it will be decided whether regulatory action is appropriate.

It is clear that the process of reaching a rational risk assessment for a substance is lengthy and laborious. There are many opportunities along the way to use interim information out of context and to make decisions which are of little benefit to human and environmental wellbeing. Such decisions may overlook the difference between **hazard** and **risk**. Classification under the DSD considers only hazard, this being an intrinsic property of a substance. Risk assessment additionally takes into account exposure of people or of the environment to the substance, recognising the realities of its production, transportation, use and disposal.

9.4 CLASSIFICATION OF PLASTICISERS

To conform with European Union law described above suppliers of plasticisers are required to give them appropriate designations (which differentiate between levels of hazard) within the various categories of hazard. Where there is no evidence of hazard no classification is required and there is no obligation to classify by analogy with chemically related substances. There is very wide disparity in the amount of data available for different plasticisers. For reasons already discussed DEHP has been the subject of toxicological research for many years and a huge body of information on this material now exists. By comparison the data available for most other plasticisers are relatively sparse. The results of the NCI/NTP study on DEHP and its inclusion in an IARC monograph and its aftermath appeared to differentiate it from other plasticisers. The possibility of many other plasticisers being classified had not arisen simply because they had not been subject to the type of testing which had been applied to DEHP.

The Seventh Amendment of the DSD (92/32/EEC implemented 31.10.93) introduced a new category of hazard to the classification rules, namely 'Toxic to Reproduction'. This requires classification of a substance according to any **known** developmental of fertility effects. However there is still no requirement for tests to be carried out for these effects on either new or existing chemicals. In fact chemicals for which sufficient data exist to allow classification for reprotoxicity are very much in the minority. A few phthalates have been tested, namely dimethyl phthalate (DMP), diethyl phthalate (DEP) dibutyl phthalate (DBP) and, not surprisingly DEHP. DMP and DEP have shown no effect and so are free of any need for classification. Rodent feeding trials involving high dietary intake of DBP and DEHP have indicated both developmental toxicity and impairment of male and female fertility. These effects had been known for years before the introduction of the classification requirement. Indeed they were taken into account when risk assessments were carried out to derive Occupational Exposure Standards which control exposure of personnel to these materials in the workplace.

With knowledge of this information, European producers of DEHP decided to classify DBP and DEHP as **Toxic to Reproduction, Category 3**. For reprotoxicity this is the lowest of the categorised levels of hazard which are ranked according to the strength of evidence for effects on humans. Category 1 includes materials for which there is evidence of causal relationship between the material and effects in humans. Materials for which there is a suspicion of human effects from studies on animals are placed in Category 2. Category 3 exists for materials where some evidence of animal effect exists but this is insufficient to presume a link with effects in humans.

As a result of classification containers holding DBP and DEHP now carry labels showing a black diagonal cross and the prescribed risk phrases:

'Possible risk of impaired fertility'

and

'Possible risk of harm to the unborn child'

together with the safety phrase:

'wear suitable protective clothing and gloves'.

Át this stage it is worthwhile recalling the intention behind the legislation which set these events in motion. It is to identify the inherent hazards of industrial chemicals in order to permit a structured assessment of the risks associated with their use so that appropriate precautions can be taken. Because there is an enormous disparity in the state of knowledge of different materials it is inevitable that some will be classified long before others.

In practice the classification requirement has produced two unintended effects. Firstly because the container labels are new and unfamiliar there is no yardstick by which the user (this may include the general public) can readily appreciate their significance and limitations. The alarming phrase 'possible risk of impaired fertility' is rather meaningless unless accompanied by some qualifying information. It has been suggested that it is analogous to labelling a car tyre with the phrase 'possible risk of explosive deflation' without any indication of safe limits on pressure or speed.

The other misunderstanding which arises relates to artificial differentiation between alternative materials for a particular application. A prime example is the use of di-n-butyl phthalate (DBP) and di-isobutyl phthalate in PVAc emulsion adhesives, this being the largest application of these C_4 phthalates. The two plasticisers are more or less technically interchangeable here and in the past selection has been made by emulsion compounders on the basis of cost/performance. Historically DBP has been subjected to far more testing for toxicity and eco-toxicity than DIBP. This is partly because of its greater commercial importance and also perhaps because it is structurally simpler and tends to be purer than the branched chain product. Suppliers have a wealth of information on which hazard classification can be allocated to DBP, resulting in the labels described above. In contrast, in the absence of evidence, positive or negative on the effects of DIBP, there is no obligation to classify it. Hence at the present time DIBP containers do not carry warning labels.

PVAc emulsions containing additives fall within the scope of the Dangerous Preparations Directive. As a result containers of PVAc adhesives etc. containing more than 5% of DBP must carry the same warning label as the plasticiser itself. This has a potential impact on a large proportion of the downstream market. Since emulsion adhesives have always been regarded as 'safe' products the presence of warning labels relating to emotional issues is something which producers would naturally prefer to avoid. Reassurance of the enormous range of downstream users about the absence of significant risk in the face of media attention and public alarm would involve a very large effort with no likelihood of success. Although containers of PVC plastisols incorporating DEHP are also subject to the same labelling requirements, their initial impact in this case has been relatively minor since their use is confined to industrial environments where the significance of the labels is better understood.

Predictably 1994 saw a large swing in demand among PVAc compounders from DBP to DIBP. This has had an effect on the availability and relative price of the latter. At the same time it is generally acknowledged that there is unlikely to be any difference between the (negligible) level of risk presented by the two plasticisers. Hence the substitution cannot be claimed to have reduced risk. It is, however, seen as a practical means of avoiding alarming labels. Some PVAc compounders have expressed interest in moving completely away from phthalates in order to avoid association with classified materials. This would involve substitution by relatively untested types of plasticiser, a move unlikely to be supported by proper risk assessment.

It appears probable that in the short term substitution of DBP in this market by alternative plasticisers will persist. In the longer term the outcome is likely to be widespread adoption of VAc/E pressure copolymers not requiring the use of external plasticisers (see Chapter 7, 7.2.1).

In the area of eco-toxicity some consideration has been given to classification of DBP under the DSD because of the existence of results indicating aquatic toxicity. It is already classified as a marine pollutant for shipping purposes on the basis of such results. The methodology for assessing the aquatic toxicity of materials of low aquatic solubility is currently being questioned by experts and it is believed that some past studies have produced misleading results. Consequently for the purposed of the DSD, DBP, like other plasticisers remains unclassified under any eco-toxicological heading. The wider subject of the environmental impact of plasticisers is discussed under 9.6.

9.5   RESTRICTIONS ON THE USE OF PLASTICISERS IN SPECIFIC APPLICATIONS

9.5.1   INTRODUCTION

The majority of Chapter 9 is concerned with issues which have a broad impact on the acceptability of plasticisers for general use. In this section we focus on three specific areas where the types of plasticisers used are restricted by toxicological considerations, namely food packaging, medical products and children's toys. Whilst these applications represent only a small proportion of plasticiser demand their obvious potential for human exposure creates a particular need for well informed risk assessment and prudence. In practice the processes by which decisions affecting the use of plasticisers in these different areas have been reached are quite diverse. In consequence the results often appear to be mutually inconsistent.

9.5.2 FOOD PACKAGING

Regulations controlling the additives which may be included in plastics used in contact with food have long been a feature of national laws in a number of European countries. These normally include a **positive list** of the permitted additives together with restrictions on their levels in the formulated plastic compound and/or the amount migrating into the food. The positive lists used in Belgium, France, Italy, the Netherlands and Spain are binding whereas those used in Germany and the U.K are optional. To overcome the barriers to trade caused by this diversity a programme of harmonisation was commenced within the EEC in 1976 with the adoption of the EC Directive 76/893/EEC. This framework directive laid down the basic principle that food contact materials should not transfer to foodstuffs any of their constituents in quantities which could endanger human health or cause a deterioration in the organoleptic characteristics of the foodstuff (i.e. cause taint). This was subsequently superseded by a second framework directive 89/109/EEC incorporating procedural arrangements designed to speed up the legislative process.

Development of European legislation relating to plastics materials and articles intended for contact with foodstuffs commenced in 1980. A series of directives has been adopted applying the principles of the framework directive to this complex area. The 'Plastics Directive', 90/128/EEC and its subsequent amendments, now incorporated into the laws of member states is a particularly important document for the packaging industry. It defines plastics materials (regenerated cellulose and elastomers are excluded and are respectively covered by separate legislation and proposals for legislation), stipulates an **overall migration limit** for plastics packaging and provides a **positive list** of monomers and other starting materials which may be used to produce food contact plastics. A further amendment of the directive to extend the positive list to additives (including plasticisers) is expected some time after 1 January 1995.

The complexities of devising a rational and comprehensive set of regulations governing plastics additives had proved so great that at the end of 1994 producers of plastics packaging were still relying on the mixed bag of established national regulations to demonstrate that they were conforming to the requirements of the framework directive. Understanding of the issues involved is assisted at this stage by a short glossary of terms commonly appearing in treatments of the subject.

Overall Migration Limit (OML)

OML is essentially an all-embracing quality standard to prevent adulteration of the food by migration of any material from the packaging. The limit has been set at 60 μg/g of food or 10 mg/dm^2 and is in force throughout the EU. The equivalence of the two values is derived from the standard model of a 1 kg cube of food with unit density and surface area 6 dm^2 totally enclosed in the packaging material. The OML represents a dietary intake of 1 mg/kg of body weight for a 60 kg person.

Tolerable Daily Intake (TDI)

TDI is the average daily intake of the substance which may be consumed over a lifetime (expressed as mg/kg of body weight) without appreciable adverse health effects. TDI values are based on toxicological data and allow large margins of safety in relation to observed no effect levels.

The Scientific Committee on Food (SCF)

SCF is a group of independent scientific experts reporting to the European Commission. Included in its responsibilities are the compilation of a positive list of additives which may be used in food contact plastics and the allocation of TDI values to such additives.

Specific Migration Limit (SML)

SML is a migration limit set for an individual chemical to ensure that it does not exceed a safe level in the packaged food. Where sufficient toxicological data exist the SML is based via a set of theoretical assumptions on the TDI for the material. It is one method of ensuring that the TDI is not exceeded and has been the approach favoured by the SCF.

Food Simulants

Food simulants are used to test packaging materials for conformance to SMLs. Migration is measured into four standard simulants representing different types of food. These are distilled water, 15% ethanol in water, 3% acetic acid in water and olive oil. The last of these generally causes the greatest migration. Because olive oil is more extractive than the fatty foods which it simulates **reduction factors** are applied in specific cases, e.g. $\div 2$ for butter and margarine.

At the time of writing the use of SMLs in the European plastics packaging legislation is far from being agreed. Whilst the approach has some advantages the enormous work load necessary for demonstrating compliance with SMLs has been pointed out and the case for considering alternative approaches has been stated.[122] The example of DEHA, the plasticiser used in largest tonnage for food packaging, has been cited to illustrate the pitfalls of using SMLs.[123] DEHA has undergone exhaustive toxicological testing on the basis of which the SCF have allocated a TDI of $0.3\,mg\,kg^{-1}$ of body weight. This would lead to an SML of $3\,mg\,dm^{-2}$ being applied to PVC cling film plasticised with DEHA. Such film currently marketed conforms to the OML of $10\,mg\,kg^{-1}$ of body weight. However migration tests have shown that it would exceed the suggested SML which, if it became mandatory would prohibit the product. Such an outcome would be at variance with the results of human exposure studies which show that in practice the current use of DEHA-plasticised film does not cause the TDI to be exceeded. In the UK a study by the MAFF produced an estimate of $0.14\,mg\,kg^{-1}$ **maximum** adult intake. This took into account the food consumption pattern shown by the UK Food Survey. The worst case assumption was made that **all** food ingested was wrapped in DEHA-plasticised film. Recently it has been shown by analysis of metabolites of DEHA in urine samples from 120 representative UK residents[124] that the mean DEHA intake is $0.045\,mg\,kg^{-1}$ (i.e. only 15% of the TDI). The relationship between DEHA intake and metabolite levels in urine had been previously established by feeding volunteers with labelled DEHA.

Alternative (and, it is argued more practical) proposals for ensuring that TDIs are not exceeded are currently under discussion by industry and regulators. It has been suggested that consumption factors should be taken into account. SMLs could then be adjusted on the basis of known use patterns for particular additives. Additionally the use of certain additives could be limited to specific polymers. Where the use of a polymer was known to be very small the requirement for an SML could then be waived. The need for measuring migration from packaging could be dispensed with altogether if industry were allowed (as it would prefer) to operate with defined composition limits for plastics products containing specific additives. This option would require a considerable initial effort to establish the relationships between additive levels in different plastics and levels of migration but thereafter compliance would be relatively simple. The outcome would then be somewhat akin to the regulatory regime administered in the USA by the internationally respected Food and Drugs Administration.

The legislative position in Europe affecting the use of plasticisers in plastics for food contact is likely to remain complex, confusing and dynamic for a considerable time. Whilst reference is made in Chapter 6, 6.8 to the major plasticisers used currently in PVC food packaging no definitive statements are attempted regarding their approval status since these could soon become dated and misleading.

9.5.3 Medical Applications

As discussed in Chapter 6, 6.9 plasticised PVC is the plastic material used in largest volume for medical products (mainly in tubing and bags for biological fluids). Until now any supplier of such products has had to demonstrate its safety to the relevant authorities in individual European countries in order to gain approval for their use. Moves towards harmonisation within the EU have resulted in the Medical Devices Directive which is expected to take effect in January 1995. In each member state various **notified bodies** have the authority to approve products according to the rules set out in the Directive. In the UK these include the BSI. The responsibility for ensuring that assessment standards are maintained lies with the **Overseeing Competent Body** for each country. In the UK this role is fulfilled by the Department of Health. Approval in any one member state will confer approval throughout the EU.

The Directive states that the approval can be based on European Pharmacopoeia monographs or, where they exist on CEN standards. The monographs generally relate to end products but in some cases are written to cover materials (e.g. plasticised PVC). In rare cases preferred compositions for such materials are stated. At the time of writing new CEN standards covering medical products are pending. When these are issued the resulting European harmonisation will extend beyond the EU. If a product does not conform to a relevant product standard or monograph then the onus is on the supplier to provide data demonstrating the safety of the product.

In contrast to the situation with food packaging there are no existing or proposed positive lists of approved plasticisers for medical products either under national or European law. Only one plasticiser is mentioned in any European Pharmacopoeia Monograph and that is DEHP. This remains by far the largest tonnage plasticiser used in medical products and has by far the most researched toxicology. The findings of the nineteen eighties have been taken into account in risk assessments confirming its approval.

This does not imply any restriction on the use of other plasticisers other than DEHP. However any supplier of a medical product containing such a plasticiser will be faced with considerable costs in generating the data needed to gain approval. Because of the commercial implications the detailed data submitted to the authorities in such cases would not necessarily become public in the scientific literature. Some PVC products containing tri-2-ethylhexyl trimellitate and butyryl tri-n-hexyl citrate have now been approved by health authorities in some European countries.

A review of the HSE aspects of using plasticised PVC in medical applications has recently been published by the European Council for Plasticisers and Intermediates.[125]

9.5.4 TOYS

Young children face particular risks from toys because of the possibility of oral ingestion of extracted material or swallowing broken off parts. There is in existence an EEC Toy Directive which provides broad guidelines on the safety factors which should be considered in the design of such products including avoidance of risks from chemical hazards. However there is in existence no positive list from which plasticisers may be selected for use in toys. Concerns have been expressed over the potential for any swallowed fragments of plasticised PVC to harden with time in the stomach as a result of plasticiser extraction, leading to a risk of physical internal injury. However attempts to simulate the postulated hardening effect in the laboratory are reported to have failed.

Heat hardenable modelling clays based on PVC plastisols are the subject of the standard EN 71 Part 5. At the time of writing this standard requires the avoidance of 'branched chain phthalates' as plasticisers in such products but allows linear phthalates and Mesamoll® (see Chapter 5, 5.13). This distinction is quite inconsistent with the approach based on risk assessment which has been applied to the development of food contact legislation. It appears to disregard the fact that the branched chain phthalates DEHP, DINP and DIDP have been far more thoroughly tested than the allowed alternatives. Moves to rationalise the situation, possibly bringing it into line with the food contact approach are being considered.

9.6   THE IMPACT OF PLASTICISERS ON THE ENVIRONMENT

The words 'environment' and 'ecology' which were not in common use thirty years ago are now included in the vocabularies of young school children and popular television presenters. Increased awareness of the damage which can be caused to eco-systems by uncontrolled industrial activity has elevated this topic on public and political agendas. To varying degrees throughout Europe the green movement is, by its activities inside and outside mainstream politics, influencing legislative priorities. For their part many industrial organisations now regard it as essential to gain green credentials by publicising the resources and effort which they spend in discharging their environmental responsibilities.

The Dangerous Substances Directive described earlier requires classification of all industrial chemicals according to their ecotoxicological (environmental) effects. Standard laboratory tests indicate the level of hazard inherent in a substance. Detailed knowledge of the way it is produced, used and disposed of is then required to assess the risks of environmental effects occurring in practice. Risk management measures can then be applied where necessary. A key step in assessing the environmental impact of a substance is the degree of difference between its **Predicted Environmental Concentration** (PEC) and its **Predicted No Effect Concentration** (PNEC). PEC can be estimated from a knowledge of how and where the substance enters the environment and how it is subsequently distributed and transferred. Alternatively actual environmental concentrations can be measured. PNEC is determined from ecotoxicological data. In rough terms the condition PEC < PNEC would indicate an absence of risk. A system which is believed to represent the state of the art for assessing the effect of chemicals on the environment has been developed by ECETOC.[126]

As a group of substances plasticisers have a combination of characteristics which have raised concerns and made them an interesting subject of environmental studies for a number of years. The volume used is large – roughly one million tonnes annually in Western Europe. They are widely spread geographically in terms of manufacture, industrial conversion processes and use of the end products incorporating them. Applications are extremely diverse and include short life and long life products used indoors and outdoors. Compounding and fabrication processes require the use of elevated temperatures at which obvious plasticiser loss occurs by volatilisation. The one feature which attracts particular attention is the fact that in the end product the plasticiser is not chemically bound to the polymer. When this concern is pursued it appears that all plasticised products in current use together with the discarded material accumulating in landfill can be considered as an enormous widely spread reservoir of plasticiser having the potential to be released into the environment. However, as discussed in Chapter 3, the loss of the commonly used plasticisers from polymers under most conditions of service is so slow as to be technically insignificant. Were this not the case then plasticised PVC would not be the useful material which it is. Hence the possibility of huge losses to the environment from flexible PVC products is far more remote than it might appear superficially to be.

With this background various studies have been carried out to estimate rates of emission of plasticisers to the environment and to measure environmental concentrations. A relatively recent report by the ECPI[127] is claimed to give a far more realistic estimate of emissions in Western Europe than reports which are sometimes quoted of earlier studies in Sweden.[128] A concise summary of ECPI's conclusions was presented at the PVC 93 conference organised by the Institute of Materials.[129] Estimates were made of total emissions of plasticisers at the various stages of their life sequence:

Manufacture → Distribution → Compounding and fabrication → Use → Disposal.

The estimates were bases on detailed knowledge of relevant technology and markets and new surveys by ECPI members of current practice in West European countries. Estimates of losses in PVC processing involved the use of DEHP as a model, i.e. the physical parameters of DEHP were taken as typical or average for plasticisers as a whole. This is believed to result in an overestimate of losses since although DEHP is the largest individual plasticiser, less volatile materials constitute the majority of tonnage of the remainder.

Table 9.1 summarises the ECPI estimates of all emissions of phthalate plasticisers for PVC.

It can be seen that about three quarters of estimated emissions are from plasticised PVC products in outdoor use. These include roof membranes, tarpaulins and car underseal. It should be emphasised that this is a very crude estimate for all outdoor applications based on the limited data available for a specific application, this being single ply roof membranes.[130] The next highest contribution, roughly an order of magnitude lower is from PVC processing. The figure of 950 t.p.a was estimated at a time when installation of effective facilities for eliminating airborne emissions in European processing plants was still only partially complete (48% for calendering of sheet and film, 75% for spread coating). Hence a reduction in this figure since then would be expected.

The ECPI conclusion that discarded PVC in landfill makes only a small contribution to total plasticiser emission may at first sight be surprising in view of the large cumulative volume involved. The estimate for Europe is based on the very low solubility of DEHP in water (<1µg/g) and pro rata on the annual combined water ingress to all landfill sites in the UK. Measurement of DEHP in effluent water from a number of these sites suggests that in practice the emissions are much lower than the 250 t.p.a. tabulated above.[131]

TABLE 9.1
ESTIMATED PHTHALATE EMISSIONS IN W. EUROPE (1990)

SOURCE	T.P.A
PRODUCTION	220
DISTRIBUTION	80
PROCESSING	
– calendered film and sheet	280
– calendered flooring	10
– spread coating	520
– other plastisols	50
– extrusion & injection moulding	90
(TOTAL PROCESSING)	950
INTERIOR END USE	
flooring-evaporation loss	20
flooring-water extraction	500*
wallcovering	20
other film, sheet, coating	40
wire, cable, profile, hose	60
(TOTAL INTERIOR USE)	640
EXTERIOR END USE	5,600*
DISPOSAL	250
TOTAL – ALL SOURCES	7,740

* Best estimates using assumptions based on limited data.
Further studies in progress.

Hence it appears that the main source of plasticisers in the environment is likely to be plasticised PVC products in outdoor use. It should be borne in mind that the material lost is subject to degradation, in particular to biological degradation. Indeed biodegradation at the PVC surface is one of the mechanisms known to contribute to the loss of plasticisers from roof membranes. In these circumstances loss from PVC gives an overestimate of emission of plasticisers to the environment.

The main purpose in gaining knowledge of emission rates is to provide a step in determining the concentrations which could be reached in the various environmental compartments – air, water, soil and sediments. Methods of arriving at these figures (including the ECETOC system mentioned earlier) take into account the rates of degradation under different conditions and the partitioning of a substance between the environmental compartments. Because of their low volatility and low water solubility (the key parameter considered here is the **octanol/water partition coefficient**) it is predicted that plasticisers will be concentrated in the organic material in sediments. Under the anaerobic conditions encountered in sediments biodegradation occurs at very low rates. It is in sediments that the highest environmental concentrations of phthalates have been measured. Any evidence of the concentration of any industrial chemical in sediments increasing with time can be a matter of concern whether or not biological effects have been found to occur.

Reliable measurement of very low phthalate concentrations in environmental samples is notoriously difficult. Because phthalates can be present as additives in so many plastics and rubber products there is considerable opportunity for contamination in handling of the samples. Failure to observe sufficiently stringent precautions to prevent this has cast doubt on some of the results reported in the past. Two recent environmental studies of phthalate levels in Sweden and Germany which applied great care to avoiding these pitfalls are likely to serve as important references in the future.

The Swedish work was carried out to allow the Ministry of the Environment to respond to a report produced by the Government appointed 'Recycling Delegation'. This report[132] had recommended a phasing out of the use of plasticised PVC by the year 2000. This was based to a considerable extent on the results of earlier measurements showing high concentrations of DEHP at some locations in Sweden. The new work was a carefully planned study involving sampling of sediments at various distances from industrial and urban areas. It showed much lower concentrations of DEHP than had been reported earlier.[133]

The study in Germany[134] produced a large volume of data showing phthalate concentrations in and around the Rhine and its tributaries in the heavily industrialised state of Nordrhein-Westfalen. It was commissioned by the responsible department of the state government. One striking finding was a decrease in phthalate levels in river sediments deposited since 1978 following progressive increases since 1945 (see Figure 9.1). Overall the conclusions of the report provide considerable reassurance that the situation is under control.

9.7 CONTROL OF PLASTICISER FUME EMISSION IN PLASTICISED PVC PROCESSING

As indicated in Table 9.1 the conversion processes having the greatest potential for release of plasticiser vapour are those having flat form with a high surface to volume ratio. At process temperatures vapour pressures of common plasticisers are sufficiently high (see Chapter 2, Figure 2.17) for evaporation to occur at a significant rate. The conditions are most pronounced in the high air flow ovens used in large scale plastisol coating operations. Production of screen printed foamed vinyl wallpaper is an extreme example.

As the plasticiser vapour generated in the process cools to ambient

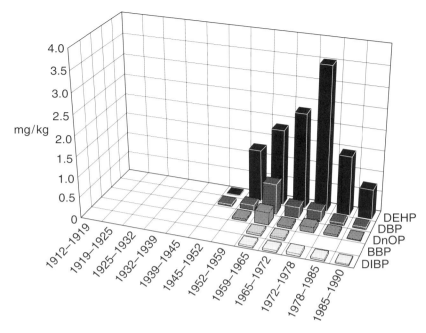

Figure 9.1 Phthalates in a dated sediment of the river Rhine.

temperature saturation vapour pressure decreases by about five orders of magnitude. Condensation occurs to produce a mist which in the absence of efficient control is emitted to atmosphere as a visible smoky plume. In extreme cases localised precipitation of the condensate has in the past resulted in noticeable greasy deposits in the vicinity of a plant with objectionable and sometimes damaging results.

In order to control this situation and to comply with any restrictions in force, plasticised PVC processes have long employed a variety of fume entrapment techniques having varying degrees of efficiency. The most popular has been the use of fibrous filters usually in the form of cylindrical 'candle' units.[135] These function by intercepting micron sized airborne droplets. Hence it is necessary for the process exhaust to be cooled to around 40°C to produce condensation to droplets prior to passage through the filter. Recovered liquid condensate is often too degraded (high odour) for reuse and is incinerated.

With the introduction of strict pollution control legislation (in the UK the 1990 Environmental Protection Act and its 1994 revision) mist filters have proved to be incapable of ensuring compliance in many cases. Fume incineration equipment has increasingly become a feature in large plants. Unit operating at 750°C or more are designed to give complete combustion of all organic airborne emissions to carbon dioxide and water. Chlorine containing components give complication which have caused some processors to reformulate when installing incinerators.

It so happens that the large scale plastisol conversion processes with the greatest potential for evaporating plasticiser (car underseal application, vinyl wallpaper and coil coating of steel) also use large quantities of volatile diluents which are completely volatilised. This has usually been the most pressing factor prompting the installation of incineration equipment. Various designs are available usually incorporating thermal recovery. Operational costs can be high unless the inputting system has been designed to minimise the dilution of the organic material by air. In addition any interruption of running can cause stresses as a result of thermal expansion and contraction. Consequently it is generally smaller operations where the greatest difficulties are encountered in adding on incinerators to existing plant. New incineration technology currently under development is expected to ease this situation.

9.8 PVC AND THE ENVIRONMENT

Since more than 90% of plasticiser consumption is in PVC, continuation of their large scale use will be very much influenced by the outcome of the ongoing debate on PVC and the environment. The issue has often been emotive and the word 'debate' with its connotations of reasoned argument is not always appropriate to the tactics employed by some protagonists.

A large section of the Green movement, through pressure groups and in several European countries through elected political representation, has been opposed for more than a decade to the continued production and use of PVC. So far industry (the producers and processors of PVC and ancillary products) have been quite unable to mollify this opposition through factual argument. There has been little constructive dialogue between the two sides which compete to influence the general public, their legislators and image conscious suppliers of consumer products (e.g. car manufacturers, supermarkets). Opposition to PVC has to be viewed in the broader context of concern over the environmental impact of chlorine and its derivatives. The tendency to automatically express opposition to all such materials has been termed 'chlorophobia' by some on the other side of the argument. The controversy surrounding chlorine and PVC has attracted popular media attention including at least one peak time television feature. The industry view was that this relied mainly on sensation and imagery rather than factual presentation. The great merits of PVC were considered to have been overlooked or trivialised.

After a slow start the PVC industry has become more actively involved in countering this negative image by bringing its own message to a wider audience. One such initiative has been the setting up in 1995 of the PVC Information Group by the European Council of Vinyl Manufacturers.[136]

PVC and plasticisers can be considered to fall into a hierarchy of environmental concerns.

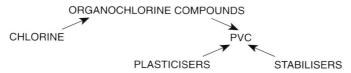

Material	Concerns
Chlorine Organochlorine compounds in general	Mercury emission from older production processes. Environmental persistence (extreme for some types). Toxic/ecotoxic effects at extremely low levels for some types including food chain effects.
PVC (the main derivative of chlorine/organochlorine chemistry)	Generation of hydrogen chloride in incineration and accidental fires. Generation of dioxins in low temperature incineration and accidental fires. Recycling difficult in some applications.
PVC ADDITIVES: – Plasticisers (large volume products essential for flexible applications which consume 1/3 of PVC produced	Toxic effects identified for some major types. Can migrate from PVC – opportunity for widespread entry into the environment. Evidence of environmental presence.
– Stabilisers (used in all applications)	Heavy metal types difficult to substitute in some PVC applications.

All of these issues are recognised and are being addressed in various ways. Activities relating to plasticisers have been the main subject of this chapter. However for anyone inclined to the view that the chlorine based industries are unacceptable and irredeemable, plasticisers have been seen as part of the edifice and hence an important target. In other words it is possible to use plasticiser issues as sticks with which to beat PVC. Conversely the publicity attached to PVC has been a major factor focusing attention on the potential risks from plasticisers.

The recommendation by the Swedish Recycling Delegation that **plasticised** PVC should be completely phased out in Sweden by the year 2000 has been mentioned already in 9.6. It was based to a considerable extent on the assertion that phthalate plasticisers were accumulating in the aquatic environment. Subsequent evidence contradicting this view was a factor contributing to the recommendation being rejected by the responsible Ministry.

Controversy over PVC has assumed a particularly high profile in Germany and at various times, in response to the pressure, local authorities have taken action to restrict its use. For example in 1993 the Parliament of the state of Hesse were reported to have upheld a ban on the use of PVC in the construction of new buildings financed with state funds.[137] At about the same time that the Recycling Delegation's report appeared in Sweden the Bundestag Enquete Kommission in Germany completed a study on the sustainable management of material streams. Its report included a set of recommendations relating to PVC.[138] The technical performance advantages of PVC and the low level of energy consumed in producing it were acknowledged as important positive attributes. The notion of the impossibility of ecologically harmless recycling and disposal of PVC was rejected. Unlike the Swedish report the official recommendations made no reference to plasticiser hazards nor did they state a need for discriminatory action against plasticised PVC. However an unofficial set of recommendations added by a minority group within the Kommission called for stronger action to substitute PVC in applications where the distribution or construction of the end product made material segregation and recycling impractical. Whilst there was no specific reference to **plasticised** PVC, wallcovering and car underseal were cited as examples of such products. These together happen to represent a significant proportion of the plasticiser market.

If the views reported above gain force then in the future the question of end product recyclability could have a considerable impact on the consumption of plasticisers in Europe. In this situation reassurances over the safety of plasticisers themselves are unlikely to be a controlling influence.

As a postscript to this chapter it is worth recalling that the discussion has focused on the HSE scene relating to plasticisers in Europe with occasional passing reference to the USA. Following current trends it is likely that in years to come the major producers and consumers of plasticised PVC products will be located in Asia and the Pacific Rim. Whilst they may be influenced by European experience and opinions it cannot be assumed that they (and their legislators) will arrive at the same conclusions on risk-benefit balance for the materials involved.

Plasticiser Manufacturers and Trade Names – 1994

This list is not claimed to be comprehensive but is believed to include the main suppliers of plasticisers to the European market at the end of 1994 together with trade names where used. The majority of these are companies with European manufacturing operations. Some are USA manufacturers exporting speciality products to Europe. For full up to date details of American plasticiser manufacturers, their products and trade names the reader is referred to *Modern Plastics Encyclopedia* which is published annually (Mcgraw-Hill Inc., 1221 Avenue of the Americas, New York, NY10020).

Because of business mergers, acquisitions and withdrawals from the market any list of this type can only be a historical snapshot and major changes over the next few years can be anticipated.

The category references are the section headings from Chapters 4, 5 and 7 where descriptions of the various plasticiser types can be found. These are grouped as follows:

4.4	dimethyl phthalate and diethyl phthalate
4.5	DBP and DIBP
4.6.1	DOP
4.6.2/3	phthalates – branched chain, other (including DINP and DIDP)
4.7	phthalates – linear
4.9.1/2	benzyl phthalates (including BBP)
4.9.4/5	phthalates – other
5.4	chlorinated paraffins
5.5	adipates, sebacates and azelates
5.6	epoxies
5.7	polyesters
5.8	phosphates
5.9	trimellitates
5.10	esters of iso- and terephthalic acids
5.11	aliphatic esters of glycols (including TXIB)
5.12	esters of polyhydric alcohols (including triacetin)
5.13	alkyl sulphonate esters

5.14   citrates
5.15   benzoates
5.16   plasticising polymers

7.4.1   hydrocarbons

X   other plasticiser types

Manufacturer	Categories	Trade names
1. Akcros Chemicals Ltd. PO Box 1, Eccles, Manchester, M30 0BH UK	5.6, 5.7	Lankroflex
2. Akzo Chemicals BV Amersfoort THE NETHERLANDS	X	Ketjenflex
3. Alusuisse Italia S.p.A. Via Vittor Pisani, 31–20124 Milano ITALY	4.5, 4.6.1 4.6.2/3, 4.7, 4.9.1/2 5.5, 5.7, 5.9	Diplast Polimix
4. Aristech Chemical Corporation Pittsburgh, PA USA	4.5, 4.6.1 4.6.2/3, 4.7, 4.9.1/2 5.5, 5.6, 5.9 X	PX-
5. Bayer AG D-W 5090 Leverkusen GERMANY	4.9.1/2, 4.9.4/5 5.5, 5.7, 5.8 5.12, 5.13, 5.16 X	Adimoll Baymod Disflamoll Mesamoll Unimoll Ultramoll Vulkanoll
6. BASF AG D-W 6700 Ludwigshafen GERMANY	4.4, 4.5, 4.6.1 4.6.2/3, 4.7 4.9.4/5 5.5, 5.7	Palatinol Palamoll
7. BUNA AG D-04212 Schkopau GERMANY	4.6.1 5.5, 5.7	Sconamoll
8. BP Chemicals Ltd Moor Lane, London EC2Y 9BU UK	4.4, 4.5 4.6.1, 4.7	Bisoflex*
9. BP Oil Hemel Hempstead HP2 4UL UK	7.4.1	Enerflex Enerpar Enerthene

Manufacturer	Categories	Trade names
10. Caffaro Industrial Auxiliaries Division 20031 Cesano Maderno (Mi) ITALY	5.4	Cloparin
11. Chemetall D-W6000, Frankfurt am Main GERMANY	7.4.1	Naftolen
12. Chemial S.p.A. Milano ITALY	4.4	Chemial
13. Chemie Linz A-4021 Linz AUSTRIA	4.5, 4.6.1 4.6.2/3, 4.7 5.5	Mollan
14. COIM S.p.A. 20019 Settimo Milanese ITALY	4.4, 4.5, 4.9.4/5 5.5	Exterplast Plaxter
15. Condea Chemie GmbH 2000 Hamburg 60 GERMANY	4.7 5.5, 5.9	Linplast Reproxal
16. Condensia Quimica Barcelona SPAIN	include 5.9	Glyplast
17. Courtaulds Chemicals Ltd. Leek ST13 8UZ UK	5.12	
18. Croda Universal Ltd. North Humberside DN14 9AA UK	4.9.4/5 5.5, 5.11, 5.14	Croda
19. DSM Resines BV 8000 AP Zwolle THE NETHERLANDS	5.7, 5.9	Synolate Uraplast
20. DuPont de Nemours International S.A. CH-1218 Geneva SWITZERLAND	5.16	Elvaloy
21. Eastman Chemical Company Kingsport, TN 37662 USA	include – 4.4 5.5, 5.9, 5.10, 5.11	Kodaflex
22. Elf-Atochem F-92091, Paris – La Defense FRANCE	4.6.1, 4.6.2/3 5.5, 5.7, 5.9	Garbeflex
23. Exxon Chemical International Marketing Inc. B-1950 Kraainem BELGIUM	4.6.1, 4.6.2 5.5, 5.9	Jayflex

Manufacturer	Categories	Trade names
24. F.M.C. Corporation (UK) Ltd Manchester M17 1WT UK	4.9.4/5 5.5, 5.7, 5.8 5.9	Kronitex Kronox Reofos Reomol Reoplex Pliabrac
25. Gas de Portugal Lisboa PORTUGAL	4.5, 4.6.1, 4.6.2/3	
26. Goodyear Chemicals Europe 91952 Les Ulis Cedex FRANCE	5.16	Chemigum
27. C. P. Hall Chicago, IL 60638 USA	include – 5.5, 5.6, 5.7	Paraplex Plasthall
28. Hatco Chemical Corp. Fords, NJ 08863 USA	include 5.9	Hatcol
29. Henkel D-W4000 Dusseldorf GERMANY	4.6.2/3, 4.9.4/5 5.5, 5.6, 5.7 X	Edenol
30. Hoechst AG D-W6230 Frankfurt am Main GERMANY	4.6.1, 4.6.2/3 5.4, 5.8	Genomol
31. Hüls AG D-W4370 Marl GERMANY	4.5, 4.6.1 4.6.2/3, 4.7 5.5, 5.8 5.9, 5.11	Nuoplaz Vestinol Witamol
32. ICI Chemicals and Polymers Ltd. Runcorn WA7 4QG UK	5.4	Cereclor
33. International Speciality Chemicals Ltd. Southampton SO4 6ZG UK	5.5, 5.9 5.11, X	Bisoflex*
34. Kemira Polymers Stockport SK12 5BR UK	5.7	Diolpate
35. Monsanto Europe S.A. B-1150 Brussels BELGIUM	4.9.1/2 5.5, 5.7, 5.8	Santicizer
36. Morflex Inc. Greensboro, NC 27403 USA	4.4 5.14	Morflex

Manufacturer	*Categories*	*Trade names*
37. Neste Oxo AB Steningsund SWEDEN	4.6.1, 4.6.2/3 4.7	Neste
38. Pantochim BELGIUM (Subsidiary of Sisas, ITALY)	4.6.1	
39. Plastificantes de Lutxana Bilbao SPAIN	4.5, 4.6.1	Indoflex
40. Sisas S.p.A Milano ITALY	4.5, 4.6.1, 4.6.2/3	
41. Thiokol Chemical Division Trenton, NJ 08607 USA	5.5 X	Thiokol
42. Traquesa 08210 Barcelona SPAIN	5.7, 5.9	Adilen
43. Union Chimique Belge S.A. Brussels BELGIUM	4.4, 4.6.1	
44. Unichema International 2802 BE Gouda THE NETHERLANDS	5.5, 5.6, 5.7	Priplast
45. Union Carbide Chemicals and Plastics Co Inc. Danbury, CT 06817–0001 USA	include – 5.5, 5.6, 5.11 X	Flexol
46. Union Camp Corp. Wayne, NJ 07470 USA	5.5, 5.7 X	Uniflex
47. Velsicol Chemical Corp. Rosemont, IL 60018 USA	5.15	Benzoflex

* The BP Chemicals trademark 'Bisoflex' is also used under license by International Speciality Chemicals.

Performance Tables for PVC Plasticisers

Whilst being far from comprehensive these tables provide comparison on a common basis of most of the significant plasticiser types covered by Chapters 4 and 5 of this book. They have the merit of being obtained from a single source providing uniformity of formulations and test methods. Thanks are due to BP Chemicals Limited for permission to reproduce information which was originally published in the form of wall charts.[139,140] It has been supplemented here by additional data produced in the same test laboratories.

Formulations (parts by weight)

For mechanical and permanence tests –

PVC suspension resin, ISO viscosity no. 125	100
Plasticiser	as shown
Dibasic lead phthalate	5
Calcium stearate	0.5

For volume resistivity –

PVC suspension resin, ISO viscosity no. 125	100
Plasticiser	as shown
Tribasic lead sulphate	5
Low molecular weight polyethylene lubricant	0.5

For plastisol viscosity –

PVC paste resin, ISO viscosity no. 130	100
Plasticiser	60

Preparation of Test Specimens

Compounds were milled for 15 minutes (5 minutes for C_4 phthalates) at 140°C (150°C for plasticisers with gelation temperature >130°C). Sheets were moulded at 178°C under a pressure of 427 kPa for 5 minutes followed by 5380 kPa for 2 minutes.

Plastisols were mixed in a Molteni vacuum mixer for 3 minutes at low speed followed by 12 minutes at high speed.

PLASTICISER	Gelation temperature °C	Plasticiser level p.h.r.	B.S softness number	Shore A hardness	100% Modulus N/mm²	Tensile strength N/mm²	Elongation at break %	Low temperature flex point °C	Volume resistivity 10^{11} Ω m	Plastisol viscosity after 1 day Pa s	Plastisol viscosity after 7 days Pa s
C₄ Phthalates											
Di-n-butyl phthalate	72	50	42	75	7.5	19.3	285	-17	12	11.8[1]	40.0[1]
Di-isobutyl phthalate	76	50	31	85	10.4	20.7	260	-2	130	9.2[1]	17.8[1]
Higher Phthalates (Branched)											
Di-2-ethylhexyl phthalate (DOP)	109	30	6	97	20.2	24.8	270	+3	220	6.9	9.1
		40	20	88	14.7	22.3	300	-10	100		
		50	34	80	10.2	20.0	325	-22	17		
		60	46	73	7.0	16.9	370	-30	4.1		
		80	69	60	3.7	13.1	420	-44	0.8		
		100	91	49	2.4	9.0	430	-51	0.4		
Di-iso-octyl phthalate	109	30	5	97	22.6	25.2	245	+9	220	6.3	7.9
		40	17	90	16.2	22.4	280	-5	100		
		50	30	82	11.4	19.7	320	-17	20		
		60	41	76	7.8	17.2	355	-26	4.5		
		80	64	63	4.5	13.0	395	-38	1.0		
		100	87	52	2.6	9.4	400	-45	0.5		
Di-isononyl phthalate (polygas-derived)	118	30	4	98	21.7	25.7	260	+8		6.9	8.3
		40	16	93	16.2	23.4	295	-7			
		50	28	84	12.2	21.0	325	-19			
		60	40	77	8.9	18.6	350	-29			
		80	64	63	5.2	14.1	390	-42			
		100	89	50	3.2	10.7	425	-46			

PLASTICISER	Gelation temperature °C	Plasticiser level p.h.r.	B.S softness number	Shore A hardness	100% Modulus N/mm²	Tensile strength N/mm²	Elongation at break %	Low temperature flex point °C	Volume resistivity 10^{11} Ω m	Plastisol viscosity after 1 day Pa s	Plastisol viscosity after 7 days Pa s
Di-3,5,5-trimethylhexyl phthalate	119	30	3	99	23.8	24.1	280	+16	450		
		40	12	93	17.2	21.5	310	0	120		
		50	23	86	12.3	19.4	330	-9	32		
		60	36	79	9.0	17.6	350	-18	10		
		80	62	64	5.5	14.2	375	-28	2.0		
		100	86	52	3.7	10.5	390	-34	0.5		
Di-isodecyl phthalate	12	30	4	98	20.4	24.4	265	+8	200		
		40	13	92	16.0	21.9	290	-8	120		
		50	23	86	12.3	19.6	320	-21	20		
		60	34	80	9.2	17.4	340	-28	5.5	8.0	9.1
		80	58	66	4.9	13.4	375	-39	1.4		
		100	83	53	3.0	10.0	405	-47	0.6		
Di-isotridecyl phthalate	146	30	3	98		26.2	265	+10	75		
		40	7	96		23.4	290	-7	15		
		50	12	93	13.0	21.0	305	-19	3.7	16.1	18.1
		60	21	87	9.7	19.0	325	-27	0.7		
		80	42	75	5.7	14.8	360	-38	0.3		
		100	66	62		12.1	385	-45			
'Linear' Phthalates											
Di-(linear C_7/C_8/C_9) phthalate	105	30	8	95	19.8	24.8	240	0	45		
		40	23	86	14.0	21.7	280	-14	5		
		50	37	78	9.7	18.6	320	-27	1.7	4.6	6.2
		60	48	72	6.6	16.2	350	-36	0.4		
		80	70	60	3.3	11.7	400	-49	0.2		
		100	93	48	2.2	8.6	430	-57			

PLASTICISER	Gelation temperature °C	Plasticiser level p.h.r.	B.S softness number	Shore A hardness	100% Modulus N/mm²	Tensile strength N/mm²	Elongation at break %	Low temperature flex point °C	Volume resistivity $10^{11}\,\Omega$ m	Plastisol viscosity after 1 day Pa s	Plastisol viscosity after 7 days Pa s
Di-(linear $C_9/C_{10}/C_{11}$) phthalate	125	30	5	97	20.4	25.0	270	0	37		
		40	14	92	14.8	22.0	290	-17	3		
		50	25	85	10.7	19.2	310	-30	0.9	3.9	4.5
		60	36	79	7.9	16.7	325	-40	0.2		
		80	59	66	4.6	12.3	360	-52	0.1		
		100	83	53	2.8	8.9	390	-60			
Diundecyl phthalate	136	30	3	98	20.7	25.9	255	-7			
		40	8	95	16.0	23.9	325	-21			
		50	17	90	12.5	20.5	350	-33	3.4	4.8	5.8
		60	28	86	9.8	12.7	365	-43			
		80	52	71	6.7	15.4	380	-54			
		100	76	59	4.7	11.2	390	-62			
Other Phthalates											
Butyl benzyl phthalate	88	30	4	98	25.2	26.2	245	+19			
		50	40	76	10.0	21.4	290	-5	40	65.0[1]	95.0[1]
		80	68	61	4.1	13.3	390	-27			
Esters of Linear Dibasic Acids											
Di-2-ethyhexyl adipate (DOA)	126	30	12	93	16.5	24.8	285	-18			
		40	24	86	10.5	21.4	325	-37			
		50	36	79	7.3	18.6	355	-52	0.3	1.8	2.3
		60	48	72	5.2	15.9	380	-60			
		80	71	59	4.2	11.7	410	-70			

PLASTICISER	Gelation temperature °C	Plasticiser level p.h.r.	B.S softness number	Shore A hardness	Modulus 100% N/mm²	Tensile strength N/mm²	Elongation at break %	Low temperature flex point °C	Volume resistivity $10^{11}\,\Omega\,m$	Plastisol viscosity after 1 day Pa s	Plastisol viscosity after 7 days Pa s
Di-(linear $C_7/C_8/C_9$) adipate	120	30	12	93	16.1	25.2	300	-22			
		40	25	85	10.8	22.3	335	-42			
		50	39	77	7.6	19.7	395	-56	0.08	1.8	2.4
		60	50	74	5.8	15.8	400	-66			
		80	71	59	4.2	11.2	430	-73			
Di-isodecyl adipate	143	30	6	97	18.3	27.2	320	-21			
		40	13	92	13.3	21.5	345	-35			
		50	21	87	10.2	18.1	350	-47	0.6	1.6	1.9
		60	32	81	8.1	15.5	355	-58			
		80	58	66	5.5	11.6	360	-69			
Di-2-ethylhexyl azelate (DOZ)		50	34	80		18.3	380	-53			
Di-2-ethylhexyl sebacate (DOS)	140	30	7		16.3	24.8	300	-17			
		50	29		8.1	18.3	345	-57			
		60	40		6.3	15.5	360	-67			
		80	64			11.4	385				
Di-2-ethylhexyl succinate		50	37	78				-45			
Polyesters											
General purpose, low viscosity modified poly (propylene adipate) MW approx 1200	130	50	22	87	15.7	21.9	290	-3	5.6	>200	>200
		60	37	78	12.2	19.9	315	-13			
		80	61	65	7.7	16.3	340	-21			

PLASTICISER	Gelation temperature °C	Plasticiser level p.h.r.	B.S softness number	Shore A hardness	Modulus 100% N/mm²	Tensile strength N/mm²	Elongation at break %	Low temperature flex point °C	Volume resistivity $10^{11}\,\Omega$ m	Plastisol viscosity after 1 day Pa s	Plastisol viscosity after 7 days Pa s
Phosphates											
Phosphate of isopropylated phenol (Synthetic equivalent of TTP)	82	30	2	99	15.3	42.9	15	>+30			
		50	23	86	6.2	22.8	270	+3	59		
		80	59	66		15.9	340	-17			
Phosphate of isopropylated phenol (Synthetic equivalent of TXP)	81	30	2	99	17.0	44.1	10	>+30		66[1]	92.5[1]
		50	17	90	6.8	22.4	290	+5			
		80	59	66		17.2	350	-15			
2-ethylhexyl diphenyl phosphate	88	30	6	97	20.9	25.7	250	+5		8.1[1]	15.8[1]
		50	38	77	10.2	19.5	260	-22	0.3		
		80	70	60	4.7	12.5	380	-45			
Trimellitates											
Tri-2-ethylhexyl trimellitate	128	30	3	98	22.7	26.3	270	+6	130	18.1	19.7
		40	10	94	17.5	23.9	300	-7	50		
		50	19	88	13.2	21.6	325	-16	8.5		
		60	31	82	9.8	19.2	345	-24	1.5		
		80	60	65	5.7	14.4	365	-37	0.6		
		100	84	53	4.2	10.8	375	-45			
Tri-(linear $C_7/C_8/C_9$) trimellitate	125	30	4	98	21.9	26.1	260	+8		9.0	9.8
		40	11	95	17.4	23.6	290	-7			
		50	24	90	13.0	21.7	330	-18			
		60	39	78	9.7	19.7	360	-28			
		80	63	65	6.3	15.0	380	-39			
		100	85	54	4.3	11.9	390	-47			

290 Plasticisers: Principles and Practice

PLASTICISER	Gelation temperature °C	Plasticiser level p.h.r.	B.S softness number	Shore A hardness	100% Modulus N/mm²	Tensile strength N/mm²	Elongation at break %	Low temperature flex point °C	Volume resistivity $10^{11}\,\Omega$ m	Plastisol viscosity after 1 day Pa s	Plastisol viscosity after 7 days Pa s
Tri-(linear C_8/C_{10}) trimellitate	135	30	3	98	20.6	23.8	270	0			
		40	7	96	16.9	21.6	290	-16			
		50	13	92	11.8	19.3	315	-28	6.7	5.7	6.2
		60	22	87	8.9	17.2	330	-38			
		80	45	73	5.2	13.1	360	-49			
		100	71	59	4.7	10.3	380	-56			
Aliphatic Esters of Glycols											
Tri (ethylene glycol) di-caprate/caprylate	138	30	12	93		26.2	295	-15.5			
		40	22	87	10.2	23.1	350	-36	0.002		
		50	33	80	7.4	20.3	380	-52			
2, 2, 4 - trimethylpentandiol 1, 3 - di-isobutyrate	123	50[2]	47					-32			
Sulphonates											
Mesamoll ®		50	33					-17			
Mesamoll II ®		50	39					-19			
Citrates											
Acetyl tri-2-ethylhexyl citrate		50	15					-22			

NOTES 1. Plasticiser level for plastisol viscosity measurement was 60 p.h.r. in all cases.
2. Test specimens for performance testing of this plasticiser were prepared by compression moulding of plastisol in order to avoid volatile loss during milling.

PLASTICISER	Plasticiser level p.h.r.	Loss on heating %	Extraction by:			
			Petrol g/m²	Mineral Oil g/m²	Olive Oil g/m²	1% Soap solution %
C₄ Phthalates						
Di-n-butyl phthalate	50	8.6	106	16	18	25
Di-isobutyl phthalate	50	10.3	75	7	9	21
Higher Phthalates (Branched)						
Di-2-ethylhexyl phthalate (DOP)	30 / 50 / 100	0.8 / 0.8 / 0.8	20 / 90 / 187	1 / 10 / 43	1 / 21 / 79	11 / 19 / 29
Di-iso-octyl phthalate	30 / 50 / 100	0.5 / 0.7 / 0.8	21 / 100 / 242	1 / 11 / 79	1.5 / 19 / 129	9 / 19 / 27
Di-isononyl phthalate (polygas-derived)	30 / 50 / 80	0.3 / 0.5 / 0.7	32 / 138 / 232	2 / 7 / 115	4 / 24 / 124	6 / 12 / 15
Di-3,5,5-trimethylhexyl phthalate	30 / 50 / 80	0.5 / 0.6 / 0.6	9.1 / 128 / 219	0.3 / 7 / 102	1 / 25 / 120	8 / 16 / 18
Di-isodecyl phthalate	30 / 50 / 80	0.3 / 0.4 / 0.7	21 / 129 / 229	3 / 10 / 148	4 / 32 / 130	2 / 9 / 10
Di-isotridecyl phthalate	30 / 50 / 80	0.1 / 0.2 / 0.2	69 / 149 / 269	11 / 48 / 115	25 / 38 / 118	0.1 / 1.5 / 6

PLASTICISER	Plasticiser level p.h.r.	Loss on heating %	Extraction by:			
			Petrol g/m²	Mineral Oil g/m²	Olive Oil g/m²	1% Soap solution %
'Linear' Phthalates						
Di-(linear $C_7/C_8/C_9$) phthalate	30	0.4	26	2	4	12
	50	0.4	116	17	26	18
	100	0.4	241	93	147	26
Di-(linear $C_9/C_{10}/C_{11}$) phthalate	30	0.2	19.2	3	6	2
	50	0.3	123	28	45	4
	100	0.4	245	191	160	5
Diundecyl phthalate	50	0.3	79	54	50	2
Other Phthalates						
Butyl benzyl phthalate	50	1.6	44	10	8	27
Esters of Linear Dibasic Acids						
Di-2-ethlhexyl adipate (DOA)	30	1.5	48	9	10	14
	50	1.9	93	48	48	27
	100	2.2	204	206	205	38
Di-(linear $C_7/C_8/C_9$) adipate	50	1.1	158	70	80	24
Di-isodecyl adipate	50	0.5	161	81	73	9
Polyesters						
General purpose, low viscosity modified poly (propylene adipate) MW approx 1200	50	0.4	29	3	4	11

PLASTICISER	Plasticiser level p.h.r.	Loss on heating %	Extraction by:			
			Petrol g/m²	Mineral Oil g/m²	Olive Oil g/m²	1% Soap solution %
Phosphates						
Phosphate of isopropylated phenol (Synthetic equivalent of TTP)	50	0.6	59	5	6	27
Phosphate of isopropylated phenol (Synthetic equivalent of TXP)	50	0.5	63	4	5	25
2-ethylhexyl diphenyl phosphate	50	0.7	97	14	18	24
Trimellitates						
Tri-2-ethylhexyl trimellitate	30 / 50 / 100	0.1 / 0.2 / 0.2	12 / 122 / 129	0.4 / 10 / 134	2 / 21 / 94	0 / 2 / 4
Tri-(linear $C_7/C_8/C_9$) trimellitate	50	0.3	107	11	26	1
Tri-(linear C_8/C_{10}) trimellitate	30 / 50 / 100	0.1 / 0.2 / 0.3	51 / 144 / 239	4 / 45 / 148	7 / 39 / 113	0.5 / 2 / 2
Aliphatic Esters of Glycols						
Tri (ethylene glycol) di-caprate/caprylate	50	3.6[1]	143	63	66	30
2, 2, 4 - trimethylpentandiol 1, 3 - di-isobutyrate	50[2]	17.5				

PLASTICISER	Plasticiser level p.h.r.	Loss on heating %	Extraction by:			
			Petrol g/m^2	Mineral Oil g/m^2	Olive Oil g/m^2	1% Soap solution %
Sulphonates						
Mesamoll ®	50	0.5				
Mesamoll II ®	50	0.3				
Citrates						
Acetyl tri-2-ethylhexyl citrate	50	0.3	128	17	30	5

NOTES 1. This figure was obtained for a typical commercial sample containing a substantial proportion of monoester.
2. Test specimens for performance testing of this plasticiser were prepared by compression moulding of plastisol in order to avoid volatile loss during milling.

Gelation temperature – a 1g sample of a 1:4 mixture of PVC suspension resin (ISO viscosity no. 125) is agitated in a glass tube containing a thermometer of slightly smaller diameter than the tube's internal diameter while the temperature is raised at a rate of 2°C per minute. The gelation temperature is the point at which rapid plasticiser absorption by the resin is observed.

Softness number – BS 2782(1970) Method 307A.

Shore A Hardness – ASTM D2240–68, 15 second indentation.

Tensile strength, elongation at break and 100% modulus – tensile strength and elongation at break BS 2782(1970) Method 301E. 100% modulus measurement was based on the same method.

Low temperature flex point – modified Clash and berg method.

Volume resistivity – minor modification of BS 2782(1970) Method 202B.

Plastisol viscosity – plastisols were stored at 23°C. At 1 day and 7 days after mixing viscosity was measured with the Brookfield viscometer model RVF at 20 r.p.m..

Loss on heating – BS 2782(1970) Method 107F. The result is expressed as % loss of compound weight.

Extraction by petrol – specimens measuring 100 × 20 × 1.27 mm were immersed in 4:1 iso-octane/toluene at 30°C for 24 hours. Absorbed 'petrol' was then evaporated at 70°C.

Extraction by mineral oil and olive oil – specimens measuring 100 × 20 × 1.27 mm were immersed in the oil at 50°C for 24 hours and wiped to remove surplus oil.

Extraction by soapy water – specimens measuring 180 × 40 × 0.15 mm were immersed in boiling 1% soap solution for 24 hours and then desiccated to constant weight. The result is expressed as % loss of *compound* weight.

References

CHAPTER 1

1. B.S. 5057 Part 1: *Specifications for water-reducing admixtures*, 1982
2. B.S. 5057 Part 3: *Superplasticising admixtures*, 1985
3. *The European Market For Plastics Additives*, Study No. 1536, Frost and Sullivan Inc., 1991
4. E. Zuhl: British Patent no. 4386, *An Improved Method of Producing a Celluloid-Like Substance*, 1901
5. E. Zuhl: British Patent no. 8072, *A New or Improved Process for the Manufacture of a Celluloid-Like Substance*, 1901
6. J.R. Darby: *Chem. Process*, July 1953
7. J.K. Sears and J.R. Darby: *The Technology of Plasticisers*, Chapters 2 and 3, John Wiley & Sons, Inc., 1982
8. D. Leuchs: *Kunststoffe*, 1956, **46**, 547
9. C. Howick: *PVC 93*, Institute of Materials Conference, Paper 38, 1993
10. G.J. Van Veersen and A.J. Meulenberg: *Kunststoffe*, 1966, **56**, 23
11. P.A. Small: *J. Appl. Chem.*, 1953, **3**, 76–80
12. D.L. Buszard: *PVC Technology*, Fourth Edition, Chapter 5, W.V. Titow ed., Elsevier Applied Science Publishers, 1984, 127
13. J.R. Darby, N.W. Touchette and J.K. Sears: *Polym. Eng. Sci.*, 1967, **7**, 295–309
14. A.C. Hecker and N.L. Perry: *SPE Ann. Tech. Conf. Prepr.*, 1960, **6**, 58.1–58.4
15. R.A. Horsley: *Progress in Plastics*, P. Morgan ed., Iliffe & Sons, 1957, 77–88
16. C. Mijrangos, C. Martinez and A. Michel: *Eur. Polym. J.*, 1986, **22**(5), 417–421
17. P. Cassagnau, M. Bert and A. Michel: *J. Vinyl Technol.*, June 1991, **13**(2), 114–119

CHAPTER 2

18. Sekisui Chem Ind., Japanese Unexamined Pat. App. no. J04068004A, 1992

19. Hayakawa Rubber, Japanese Unexamined Pat. App. no. JP04106148A, 1992
20. H. Bevan and A.S. Wilson: BP Chemicals unpublished work
21. J.T. Renshaw and J.A. Cannon: *Polym. Eng. Sci.*, 1974, **14**, 473–477
22. *Tinuvin 541 Liquid Light Stabiliser for Polyvinyl Chloride*, Ciba Geigy Ltd., 1–1/92.
23. C.E. Anagnostopolous, A.Y. Coran and H.R. Gamrath: *Modern Plastics*, October 1965, 141–144 and 187
24. H. Luther, F.O. Glander and E. Schleese: *Kunststoffe*, January 1962, 52, 7–14
25. M.V. Patel and M. Gilbert: *Plastics and Rubber Processing and Applications*, 1987, **8**, 215–217
26. J.P. Tordella: *J. Appl. Phys.*, 1956, **27**, 454–458
27. E.H. Foakes: *British Plastics*, February 1966, 92–95
28. L. Mentovay: *Paint and Varnish Production*, March 1974, 31–36
29. P.R. Graham and J.R. Darby: *SPE Journal*, January 1961, 91–95
30. J.A. Greenhoe: *Plastics Technology*, October 1960, 43–47
31. A.C. Poppe: *Kunststoffe*, 1982, **72**(1), 13–16
32. K.L. Hoy: J. *Appl. Pol. Sci.*, 1966, **10**, 1871–1891
33. *Plasticisers in the Plastics Industry*, Huls Aktiengesellschaft, Oct 1992, 52–53
34. S.H. Pinner: *Plastics*, May 1965, **30**, 82–85, and June 1965, 220–226
35. Plasticisers Range Chart M5989e, BASF Aktiengesellschaft, September 1991
36. M. Gilbert: *J.M.S.-Rev. Macromol. Chem. Phys.*, Marcel Dekker, Inc., N.Y. (1994), **C34**(1), 77–135
37. Booklet series, *Cereclor in PVC Compounds*, ICI Chemicals and Polymers Ltd.
38. O.K. Barashkov and R.S. Barstein: *Plasticheskie Massy*, 1985, **11**, 19
39. P.G. Clutterbuck and A.S. Wilson: BP Chemicals unpublished work.
40. J.K. Sears and J.R. Darby: *The Technology of Plasticisers*, John Wiley and Sons Inc., 1982, 392–401
41. J.K. Sears and J.R. Darby: *The Technology of Plasticisers*, John Wiley and Sons Inc., 1982, Chapters 4 and 5
42. K. von Leilich: *Kolloid-Z*, 1942, **99**, 107–113

CHAPTER 3

43. J.R. Darby, N.W. Touchette, W.M. Rinehart and T.C. Mathis: *The Technology of Plasticisers*, John Wiley and Sons Inc., 1982, 544–545
44. M. Royen: *A.S.T.M. Bulletin*, June 1960, 43

45. D.M. Pugh and B.I.D. Davis: 'High Temperature Plasticisers for PVC', *Plastics*, February and March 1965

46. H. Domininghaus: *Plastverarbeiter*, 1981, **32**, 84

47. D. Jackowski and A.C. Poppe: *Kunststoffe*, 1992, **82**(9), 52

48. Rover Group: Private communication with A.S. Wilson, 1994

49. C. Howick: *Materials World*, November 1994, **2**(11)

50. S.H. Pinner and B.H. Massey: *British Plastics*, October 1963, 574

51. M.C. Reed, H.F. Klemm and E.F. Schulz: *Ind. Eng. Chem.*, 1954, **46**, 1344–1349

52. *Bisoflex Trimellitates in Non-Marring PVC Compounds*, Technical Service Note 4, International Speciality Chemicals Ltd., August 1993

53. D.F. Cadogan and D.J. McBriar: *PVC 87*, Paper 22, Plastics and Rubber Institute Conference, April 1987

54. Nippon Oils & Fats, Japanese Unexamined Patent Application no. J62275136A, 1987

55. Yeda Research and Development Co. Ltd., US Patent no. 4806393, 1989

56. Mitsubishi Kasei Vinyl, Japanese Unexamined Pat App no. J01156060A, 1989

57. A. Senoune and J.M. Vergnaud: *Eur. Pol. J.*, 1992, **28**(3), 311–314

58. Mitsubishi Monsanto, Japanese Unexamined Patent Application no. J62037159A, 1987

59. Y. Iriyama and H. Yasuda: Applied Polymer Symposium 1987, Publ. *J. Appl. Pol. Sci.*, 1988, **42**, 97–124

60. Solvay & Cie, French Patent no. 2597788, 1987

61. Sumitomo Electric Industries, Japanese Examined Patent Application no. J94025271B2, 1994

62. E.N. Zil'berman, O.D Strizhakov and E.M. Perepletchivoa: *Sov. Plast.*, December 1966, **12**, 31–34

63. I.S. Biggin, D.L. Gerrard and G.E. Williams: *Journal of Vinyl Technology*, December 1982, **4**(4)

64. D.L. Gerrard, I.S. Biggin and H.J. Bowley: *PVC 87*, Paper 3, Plastics and Rubber Institute Conference, April 1987

65. J. Stepek, Z. Vymazal and B. Dolezel: *Modern Plastics*, 1963, **40**(10), 146–147 & 205

66. R.I. Sloss: *Special. Chem.*, July/Aug 1992, **12**(4), 257–258

67. D.L. Buszard: *PVC 93: The Future*, Paper 45, Institute of Materials Conference, April 1993

68. P. Carty and S. White, *Polymer*, 1992, **33**(5), 1111–1112

CHAPTER 4

69. USS Corp., U.S. Patent no. 3354176, 1967

70. J. Hannay: *The Grower*, 1980, **21**, 28
71. Hardwick R.C. *et al: Ann. Appl. Biol.*, 1984, 97
72. *Jayflex Plasticisers – Phthalates and Adipates*, Exxon Chemical International Marketing
73. *European Chemical News*, 27 July 1994, 26
74. *Polymer Modifiers*, Eighth Edition, Monsanto Europe S.A., 1992
75. Denki Kagaku Kogyo KK, Japanese Unexamined Pat. App. no. J 62158734A, 1987

CHAPTER 5

76. K.L. Houghton: *Kirk-Othmer Encyclopedia of Chemical Technology*, Fourth Edition, Vol. 4, John Wiley & Sons Inc.
77. Booklet, *Cereclor in PVC Compounds – Compatibility Graphs*, ICI Chemicals and Polymers Ltd.
78. *Plasticisers in the Plastics Industry*, Huls Aktiengesellschaft, October 1992
79. J. Fath: *Modern Plastics*, 1960, **38**(8), 135
80. D.L. Buszard: 'Plasticisers for PVC – A Review', *PVC Processing II*, Plastics and Rubber Institute Conference, 1983
81. J. Moseley and P. Dawkins: *Chem. Ind.*, 1978, **16**, 620
82. S.G. Shankwalkar and D.G. Placek: *Ind. Eng. Chem. Res.*, 1992, **31**(7), 1810–1813
83. L.A. Dupont, V.P. Gupta: *J. Vinyl Technol.*, June 1993, **15**(2), 100–104
84. BP Chemicals Ltd., Unpublished information
85. Mesamoll technical information sheet 1.1.1.1, Bayer AG
86. S. Northrup: Private Communication Baxter Health Care USA to ECPI, cited in *Medical Applications of Plasticised PVC*, ECPI Review, CEFIC, Brussels, 1994

CHAPTER 6

87. M.A. Barnes: *PVC 93 – The Future*, Paper 51, Institute of Materials Conference, 1993
88. G. Pastuska: *Kaut.u.Gummi Kunst.*, Oct 1983, **36**(10), 862–865
89. A.S. Wilson, D.L. Gerrard and H.J. Bowley: *PVC 87*, Paper 8, Plastics and Rubber Institute Conference, 1987
90. R.B. Pearson: *PVC 93 – The Future*, Paper 6, Institute of Materials Conference, 1993
91. C. Blass: PVC *93 – The Future*, Paper 42, Institute of Materials Conference, 1993

92. European Council for Plasticisers and Intermediates (ECPI) Report, *Medical Applications of Plasticised PVC*, CEFIC, Brussels, 1994
93. B.A. Myrhe *et al: Vox. Sang.*, 1985, **48**, 150–155
94. IBR Project 0–0–106–87, *in vitro Simulation der Magen-Darm-Passage mit Fimo-Modelliermasse*
95. H.C. Bowley, D.L. Gerrard and I.S. Biggin: *Journal of Vinyl Technology*, **10**, 50–52
96. I.S. Biggin and D.L Gerrard: BP Chemicals unpublished work

CHAPTER 7

97. N. Hazell: 'Water Based Coatings: A New Concept for Coalescing Solvents in Latex/Dispersion-Based Coatings', *J.O.C.C.A.*, (in the press)
98. Technical information sheet: *Bisoflex DBP and DIBP in PVA Emulsions*, BP Chemicals Ltd.
99. Technical information sheet: *Reoplex 400*, FMC Corporation (UK) Ltd., 1993
100. *Chemical Economics Handbook*, SRI International, 1990
101. Monsanto, U.S. Patent no. 5137954
102. J.A. Brysdon: *Plastics Materials*, Iliffe Books, 1966
103. Technical Service Note 5: *Bisoflex 111 – Optimum Plasticiser for Synthetic Rubbers*, International Speciality Chemicals Ltd., October 1993
104. W.R. Grace: European Patent Application EP497501A1, 1992

CHAPTER 8

105. R. Caulcutt: *Achieving Quality Improvement*, Chapman and Hall
106. W.E. Deming: *The New Economics*, MIT
107. DIN 55 350, Parts 18 4.1.1 and 18 4.1.2
108. J.K. Sears and J.R. Darby: *S.P.E. Journal*, 1962, **18**, 671–677
109. ERA Technology Ltd., Cleeve Road, Leatherhead, Surrey, KT22 7SA, UK
110. *Plasticisers in the Plastics Industry*, Hüls Aktiengesellschaft, October 1992, 50–51
111. Palatinol AH Technical Leaflet M2189e, BASF Aktiengesellschaft, September 1989
112. W. Freitag: 'Analysis of Additives', Chapter 20, *Plastics Additives*, Fourth Edition, Gächter and Müller eds, Hanser Publishers, 1993

113. I.S. Biggin, S. Jones and P. Wusteman: BP Chemicals unpublished work

114. T.P. Hunt, C.J. Dowle and G. Greenway: *Analyst*, December 1991, **116**(199), 1304

115. P.B. Mansfield: *Chemistry and Industry*, 10th July 1971, 792–873

CHAPTER 9

116. *IARC Monographs*, 1982, **29**

117. 'Di-2–ethylhexyl phthalate', *BUA Substance Report* 4, 1986

118. *Official Journal of the European Communities*, 7.8.90, Commission Decision 90/420/EEC

119. D.M.P. Pugh: 'Plasticisers-Nasty or Nice', *PVC 93*, PRI Conference, April 1990

120. *How to Comply with Proposition 65: DEHP Exposure*, Chemical Manufacturers Association, March 1985

121. *Croner's Hazard Information and Packaging*, September 1994, Croner Publications, Kingston-on-Thames, UK

122. Consultation Document FCN 723, Ministry of Agriculture, Fisheries and Food: *Deregulation Study on Food Contact Materials*, April 1994

123. D.F. Cadogan: Presentation to the 3rd Annual Seminar of the International Bar Association, Brussels, November 1993

124. N.J. Loftus, B.H. Woollen, G.T. Steel, M.F. Wilks and L. Castle: *Fd Chem. Toxic.*, 1994, **32**(1), 1–5

125. European Council for Plasticisers and Intermediates (ECPI) Report: *Medical Applications of Plasticised PVC*, CEFIC, Brussels, 1994

126. *Technical Report No. 51*, European Centre for Ecotoxicology and Toxicology of Chemicals, January 1993

127. *Assessment of the Release, Occurrence and Possible Effects of Plasticisers in the Environment: Phthalate Esters used in PVC*, ECPI Report, 1993

128. A. Thuren: 'Determination of phthalates in aquatic sediments', *Bull. Environ. Contam. Toxicol.*, 1986, **36**, 33–40

129. D.F. Cadogan, M. Papez, A.C. Poppe, D.M. Pugh and J. Scheubel: *PVC 93*, Paper 32, Institute of Materials Conference, April 1993

130. G. Patuska, W. Kerner-Gang and U. Just: *Kaut.u.Gummi Kunst.*, May 1988, **41**(5), 451–454

131. Aspinwall & Co.: 1991–1992, Private communication with ECPI

132. *PVC – a plan for the avoidance of environmental effects*, Chapters 15–17, The Recycling Delegation's review of PVC on behalf of the Swedish Government, June 1994

133. *Phthalates in Sediments*, IVL The Environmental Research Institute, Report No. 1167, 1994,

134. K. Furtmann: Dissertation, Universitat-Gesamthochschule Duisburg, 1993, Reprinted in *LWA-Materials* 6, 1993, Landesumweltamt NRW, Box 10 23 63, Essen, Germany
135. *Modern Plastics International*, June 1971, 54–55
136. *Plastics and Rubber Weekly*, (lead news item), 9/29/94
137. *Plastics and Rubber Weekly*, (in lead news item), 17/07/93
138. Bundestag Enquete Kommision: *The Products of Industrial Society – Perspectives on Sustainable Management of Material Streams*, (Federal), 12/07/94

APPENDIX 2

139. Wallchart: *Bisoflex Plasticisers for PVC*, BP Chemicals Ltd, July 1977
140. Wallchart: *Bisoflex Plasticisers for PVC*, BP Chemicals Ltd, June 1990

Index